Lecture Notes in Mathematics

Edited by A. Dold and B. Eckmann

568

Robert E. Gaines
Jean L. Mawhin

Coincidence Degree, and Nonlinear Differential Equations

Springer-Verlag
Berlin · Heidelberg · New York 1977

Authors
Robert E. Gaines
Colorado State University
Department of Mathematics
Fort Collins
Colorado 80523/USA

Jean L. Mawhin
Université Catholique de Louvain
Institut Mathématique
B-1348 Louvain-la-Neuve/Belgium

Library of Congress Cataloging in Publication Data

Gaines, Robert E 1941-
Coincidence degree, and non-linear differential equations.

(Lecture notes in mathematics ; 568)
Includes bibliographical references and index.
1. Differential equations, Nonlinear. 2. Boundary value problems. 3. Coincidence theory (Mathematics) I. Mawhin, J., joint author. II. Title. III. Series: Lecture notes in mathematics (Berlin); 568.
QA3.L28 no. 568 [QA372] 510'.8s [515'.35] 76-58453

AMS Subject Classifications (1970): 34B15, 34K10, 35J65, 47H15, 55C20

ISBN 3-540-08067-8 Springer-Verlag Berlin · Heidelberg · New York
ISBN 0-387-08067-8 Springer-Verlag New York · Heidelberg · Berlin

Printing and binding: Beltz Offsetdruck, Hemsbach/Bergstr. 2141/3140/543210

To Margaret, Marie, Valérie, Jean
and
Martha, Laura, Elissa.

TABLE OF CONTENTS

I. INTRODUCTION

This work has its origin in lectures given by J. Mawhin in 1974 at the University of Brasilia and by R.E. Gaines in 1975 at the University of Louvain. Those lectures respectively covered chapters II to IV, VII to IX and chapters V-VI. Chapters X to XII have been added to include more recent material.

The emphasis of the work is on the use of topological degree techniques in studying alternative problems, i.e. problems which can be written as operator equations of the form

$$Lx = Nx \tag{1.1}$$

in a suitable abstract space, with L linear and non-invertible. As shown in chapter II many techniques have been developed to handle equations of the form (1.1) and research in this field is still very active.
A principal aim of these lecture notes is to show that by proving once and for all, for coupled mappings (L,N) satisfying certain conditions, a number of properties quite similar to those of Leray-Schauder degree, one is able to study many problems of type (1.1) in an unified way.
This is the so-called coincidence degree theory which is described in chapters III and leads in chapter IV to general useful coincidence theorems for L and N, i.e. existence theorems for (1.1) and in particular to a continuation theorem of Leray-Schauder type.

The applicability of those theorems, as in any degree theory, depends upon the obtention of a priori bounds for the solutions of the equation. Chapter V consider the problem of a priori bounds in the case of boundary value problems, including periodic solutions, for ordinary differential equations. The emphasis is placed on the determination of a priori estimates through the geometric properties of the vector field defined by the differential equation. The covered material, which includes lower and upper solutions, differential inequalities, Nagumo conditions, Gustafson-Schmitt-Bebernes convexity conditions, Krasnosel'skii's guiding functions, ... is generally classical but the approach is original in several places. In chapter VI we consider the problem of the approximation of the solutions of (1.1) by Galerkin-type methods when existence has been proved by the techniques of Chapter IV. The specific case of linear boundary value problems for nonlinear ordinary differential equations is treated in detail.

In chapter VII we return to abstract equations (1.1) where N grows at most at a linear rate in x. A unified treatment is given of generalized versions of recent results in this domain due to Cronin, De Figueiredo, Fabry, Franchetti, Fučik, Kučera, Nečas, ... Those results are applied to semilinear elliptic partial differential equations in chapter VIII which covers in a systematic way most of the recent contributions in the line of Landesman and Lazer's pioneering work. In chapter IX we study periodic solutions for ordinary and functional differential equations emphasizing nonlinearities with a growth at most linear. However, in the case of functional differential equations, techniques reminiscent of the ones of chapter V are also considered.

Chapter X is a short description of the use of coincidence degree in bifurcation theory. One obtains in a more general setting a local Krasnosel'skii type theory which facilitates corresponding extensions of the Rabinowitz global results. In chapter XI we briefly describe Hetzer's extension of coincidence degree to the case of nonlinear perturbations having k-set contraction properties instead of compactness. Applications are given to generalizations of results of Kacurovskii, Petryshyn and Amann. When N is Lipschitzian with a sufficiently small Lipschitz constant, which is the situation of classical Liapunov-Schmidt method, one then relates the coincidence degree of L and N to the Brouwer degree of the mapping associated with the classical bifurcation equation.

In all the above chapters we essentially assume that the Fredholm index of L is zero. Chapter XII treats cases where the Fredholm index of L is strictly positive in the line of the recent work of Nirenberg, Rabinowitz and Schechter.

Each of these chapters is followed by bibliographical notes refering to the original papers and giving sugestions for further reading. All the references indicated in those bibliographical notes are given in a more complete setting in the list of references at the end of the volume. This list moreover contains references to recent papers which are not explicitely described in this work but which are close in spirit or results. The chapter(s) to which those papers are related is then indicated.

In preparing this set of notes we have tried, after developing the general theory, to take our examples more in related papers appearing in the literature then in our own work. Most of our work is already written in the spirit of this monograph so that we have found a duplication unnecessary. In this way we hope to have facilitated the access to this part of nonlinear functional analysis and nonlinear boundary value problems. Our expectations will be fulfilled if this work can suggest further research in the wide area of alternative problems, nonlinear differential equations and applications to science and engineering.

II. ALTERNATIVE PROBLEMS : AN HISTORICAL PERSPECTIVE

1. The study of differential or integral or other equations which, written in operator form, are of the type

$$Lx = Nx \qquad (II.1)$$

with L (resp. N) a linear (resp. nonlinear) mapping between some topological vector spaces X and Z, and with L non invertible seems to have been initiated by Lyapunov (Zap. Akad. Nauk. St. Petersbourg (1906, 1908, 1912, 1914) in his study of integral equations related to a problem of equilibrium of rotating fluids, and by E. Schmidt (Math. Ann. 65 (1908) 370-399) in his theoretical work on nonlinear integral equations. Written in abstract setting their method consists basically to write (II.1) in the equivalent form

$$Lx = (I - Q)Nx, \quad QNx = 0$$

(notations are those of Chapter III for the mappings associated to L)
or

$$x - Px = K_{P,Q}Nx, \quad QNx = 0 .$$

Now if we set

$$Px = y, \quad (I - P)x = z$$

we obtain the equivalent system

$$z = K_{P,Q}N(y+z), \quad QN(y+z) = 0. \qquad (II.2)$$

Now if some "smallness" and "regularity" conditions are assumed for N, the first equation in (II.2), considered like an equation in z depending upon the parameter y, will be solved using Banach fixed point theorem or implicit function theorem to give a solution z(y) which will depend "regularly" upon y. Hence the solution of (II.2) will be reduced to the solution of equation (in y)

$$QN[y + z(y)] = 0, \qquad (II.3)$$

usually known as bifurcation or branching or determining equation. The interest of the method is that the spaces in which the left hand member of (II.3) is defined and take values have been made "smaller" (often one goes from infinite into finite dimensional spaces). But, due to the presence of the term z(y), (II.3) is not usually known in explicit form. This methods is generally known as Lyapunov-Schmidt method and diversification occurs in the way of getting and of studying the bifurcation equations. Let us quote in this line the basic works of Cesari (Atti Accad. Mem. Cl. Fis. Mat.Nat. (6) 11(1940) 633-692), Caccioppoli (Atti Accad. Naz.Lincei Rend. Cl. Sci. Fis. Mat. Natur. 24(1936) 258-268, 416-421), Shimizu (Math. Japan. 1 (1948) 36-40) Cronin (Trans. Amer. Math. Soc. 69 (1950) 208-231), Bartle (Trans. Amer. Math. Soc. 75 (1953) 366-384), Hale (Riv. Mat. Univ. Parma 5 (1954) 281-311) Lewis (Ann. of Math. 63 (1956) 535-548), Vainberg and Trenogin (Russ. Math. Surveys 17 (1962) 1-60), Antosiewicz (Pacif. J. Math. 17 (1966), 191-197).

Expositions can be found in the books or lecture notes of Friedrichs ("Special Topics in Analysis", Lect. Notes, New York University, 1953-54), Nirenberg ("Functional Analysis", Lect. Notes, New York University, 1960-61), Krasnoselskii et al. ("Topological Methods in the Theory of Nonlinear Integral Equations", Pergamon, 1963, "Approximate Solutions of Operator Equations", Noordhoff, 1971), Vainberg and Aizengendler ("Progress in Math.", vol. II, Plenum, 1968), Hale ("Ordinary Differential Equations", Wiley, 1969; "Applications of Alternative Problems", Brown University Lect. Notes 71-1, 1971), Mawhin ("Equations non-linéaires dans les espaces de Banach", Rapp. Sém. Math. Univ. Louvain n° 39, 1971).

All the above quoted papers (except the notes by Nirenberg, Hale and Mawhin) correspond to problems in which the bifurcation equations are finite-dimensional. The first solved example leading to infinite-dimensional bifurcation equations seems to be the problem of periodic solutions of some weakly nonlinear hyperbolic equations initiated by Cesari ("Nonlinear differential equations and nonlinear mechanics", 1963, Academic Press, 33-57). An account and a bibliography about this work can be found in the lectures notes of Hale quoted above.

2. This problem of perturbed hyperbolic equations has led Rabinowitz (Comm. Pure Appl. Math. 20(1967), 145-206) more than fifty years after Lyapunov and Schmidt, to an alternate way of considering (II.2). First solve the second equation in (II.2) considered as an equation in y with z as parameter. Introduce then the solution y(z), supposed to be sufficiently "regular" in z, in the first equation of (0.2). Then solve the resulting equation

$$z = K_{P,Q} N[y(z) + z] \qquad (II.4)$$

using Banach or Schauder fixed point theorem. Usualy, monotone operator theory is used to solve the first part of the problem and interesting results using this approach have been obtained in the theory of periodic solutions of weakly nonlinear equations by Hall (J. Differential Equ. 7(1970) 509-526; Arch. Rat. Mech. Anal. 39 (1970), 294-332), de Simon and Torelli (Rend. Sem. Mat. Univ. Padova 40(1968), 380-401) and the method has been theoretically improved by Hall (Trans. Am. Math. Soc. 161(1971) 207-218) and Sova (Comment. Math. Univ. Carolin. 1972). See also the work of Klingelhofer (Arch. Rat. Mech. Anal. 1970) on some elliptic equations.

3. A third possible approach for system (II.2) is to try to solve the two equations simultaneously. A corresponding iterative process has been introduced, for periodic solutions of ordinary differential equations, by Banfi (Atti Accad. Sci. Torino

Cl. Sci. Fis. Mat. Natur. 100(1966), 471-479) and independently by Lazer (SIAM J. Appl. Math. 15(1967) 1158-1170). Their work has been put in an abstract setting by Fabry (Boll. Un. Mat. Ital. (4)4(1971) 687-700) and Hale (see the above quoted Brown Univ. Lect. Notes, 1971). A more direct approach has been given by Mawhin("Equa-Diff 70", C.N.R.S., Marseille, 1970; see also the above quoted Lecture Notes) and systematically developed in the book, by Rouche and Mawhin, "Equations différentielles ordinaires", vol. II, ch. X, Masson, 1973.

4. All the above discussion refers to a "small" nonlinearity N in (II.1). It was the merit of Cesari (Contributions to Differ. Equ. I(1963) 149-187, Michigan Math. J. 11(1964) 385-418) to show that a finite-dimensional bifurcation equation analogous to Lyapunov-Schmidt ones could be associated to a large class of equations of type (II.1) with large nonlinearity. Although the dimension of the system of bifurcation equations obtained in this case is usually larger than the dimension of the null space of L, this method was successfully applied in a number of difficult problems and also developed theoretically by Cesari himself and Knobloch (Michigan Math. J. 10 (1963) 417-430; Math. Z. (1963)82, 177-197), Locker (Trans. Amer. Math. Soc. 128(1967) 403-413), Bancroft, Hale and Sweet (J. Differential Equations 4(1968)177-202), Williams (Michigan Math. J. 15(1968) 441-448), Harris, Sibuya and Weinberg (Arch. Rat. Mech. Anal. 35(1969) 245-248, SIAM Studies in Appl. Math. 5, 1969, 184-187), Zabreiko and Strygina (Matem. Zam. 9(1971) 651-662). Incidentally this method furnished in some cases a theoretical justification of the Galerkin's method of approximation. Also by coupling Cesari's method with the knowledge of a priori estimates for the solutions, Mawhin (Bull. Soc. R. Sci. Liège 38(1969) 308-398) introduced for the first time an existence theorem for periodic solutions of differential equations which was the prototype of the extended Leray-Schauder continuation theorem developed in the frame of coincidence degree theory considered further. A new idea was also introduced by Gustafson and Sather (Arch. Rat. Mech. Anal. 48(1972) 109-122) who showed that in some cases the use of monotone operator theory could avantageously replace the use of Banach fixed point theorem in solving

the equation corresponding to the first one of (II.2) in Cesari's approach. Also an interesting approach related to the ideas of this section and using approximate solutions was devised by Sweet (Math. Syst. Theory, 1970). A survey of this material with a more complete bibliography has been given by Hale (see the above quoted Brown Univ. Lecture Notes) and by Cesari (in "Nonlinear Mechanics", CIME, 1972).

5. To our knowledge there is no work devoted to an extension of the approach devised by Rabinowitz, Hall, ... (cf. section 2) to equations with large nonlinearities. Maybe a possibility in this way could be to replace the use of Banach or Schauder fixed point theorem for solving (II.4) by topological degree and a priori estimates of the solutions. Such an approach could lead to interesting results for nonlinear hyperbolic equations.

6. The first result in extending the third approach considered in section 3 to large nonlinearities is due to Lazer (J. Math. Anal. Appl. 21(1968) 421-425) and corresponds to periodic solutions of a particular second order equation. See also Lazer and Leach (Ann. Mat. Pura Apll .(4) 82(1969) 49-68). Lazer's approach, which makes an clever use of Schauder's fixed point theorem, was also used for some semi-linear Dirichlet problems (Landesman and Lazer, J. Math. Mech. 19(1970) 609-623. In all those papers, the nonlinearity N is not small but however $\|Nx\|/\|x\| \to 0$ if $\|x\| \to \infty$. By noting that, in the problem of periodic solutions of ordinary differential equations, the system (II2), and hence equation (II1) was equivalent to an unique fixed point problem, essentially equivalent to Lazer's one, of the form

$$x = Px + (JQ + K_{P,Q})Nx \qquad \text{(II.5)}$$

where $J : \operatorname{Im} Q \to \ker L$ is any isomorphism (a special case was also used independently as intermediate tool in Zabreiko and Strygina's paper on Cesari's method quoted above) and by applying

Leray-Schauder degree to (II.5), Mawhin (Bull. Acad. R. Belg. Cl. Sci. (5) 55(1969)934-947) proved a number of existence theorems. In particular he gave a simpler proof of his generalized continuation theorem mentioned in section 4, and extended it successively to functional differential equations (Mawhin, J. Differential Equations 10(1971) 240-261) and to operator equations (Mawhin, lecture notes quoted above). As it was also emphasized by the independent work of Nirenberg ("Trois. Coll. CBRM d'Analyse Fonctionnelle", Vander, 1971, 57-74) generalizing Landesman and Lazer's paper on Dirichlet problem, Mawhin's work showed that topological degree was more adapted to the study of (II.5) than Schauder's fixed point theorem and considered many different kinds of nonlinearities. Then the introduction of coincidence degree theory (Mawhin, J. Differential Equations 12(1972) 610-636) for some nonlinear perturbations of Fredholm mappings, showed that the generalization in this frame of Leray-Schauder continuation theorem (Leray-Schauder, Ann. Ec. Norm. Sup. (3) 51(1934) 45-78) could serve as underlying principle for the problems described in this section and for many other results obtained from different arguments. This will be the subject of the next chapters and more complete bibliographical notes will be given there.

III. COINCIDENCE DEGREE FOR PERTURBATIONS OF FREDHOLM MAPPINGS

A. Algebraic preliminaries

1. Let X, Z be vector spaces, dom L a vector subspace of X and

$$L : \text{dom} L \subset X \to Z$$

a linear mapping. Its kernel $L^{-1}(0)$ will be denoted by Ker L and its range L(dom L) by Im L.

Let $P : X \to X$, $Q : Z \to Z$ be algebraic projectors (i.e. linear and idempotent linear operators) such that the following sequence is exact :

$$X \xrightarrow{P} \text{dom } L \xrightarrow{L} Z \xrightarrow{Q} Z \tag{III.1}$$

(which means that

$$\text{Im } P = \text{Ker } L \text{ and } \text{Im } L = \text{Ker } Q.)$$

If we define

$$L_P : \text{dom } L \cap \text{Ker } P \to \text{Im } L$$

as the restriction $L|\text{dom } L \cap \text{Ker } P$ of L to dom L $\cap$ ker P, then it is clear that L_P is an algebraic isomorphism and we shall define

$$K_P : \text{Im } L \to \text{dom } L$$

by

$$K_P = L_P^{-1} .$$

Clearly, K_P is one-to-one and

$$PK_P = 0. \tag{III.2}$$

Therefore, on Im L,

$$LK_P = L(I-P)K_P = L_P(I-P)\ K_P = L_PK_P = I \tag{III.3}$$

and, on dom L,

$$K_PL = K_PL(I - P) = K_PL_P(I - P) = I - P. \tag{III.4}$$

2. Now let

$$\text{Coker } L = Z/\text{Im } L$$

be the quotient space of Z under the equivalence relation

$$z \sim z' \Longleftrightarrow z - z' \in \operatorname{Im} L.$$

Thus, Coker L = $\{z + \operatorname{Im} L : z \in Z\}$ and we shall denote by $\Pi : Z \to \operatorname{Coker} L$, $z \longmapsto z + \operatorname{Im} L$ the canonical surjection. Clearly,

$$Qz = 0 \Longleftrightarrow z \in \operatorname{Im} L \Longleftrightarrow \Pi z = 0. \qquad \text{(III.5)}$$

3. Proposition III.0. If there exists a one-to-one linear mapping

$$\Lambda : \operatorname{Coker} L \to \operatorname{Ker} L$$

then equation

$$Lx = y \ , \quad y \in Z$$

is equivalent to equation

$$(I-P)x = (\Lambda\Pi + K_{P,Q})y$$

where $K_{P,Q} : Z \to X$ is defined by

$$K_{P,Q} = K_P(I - Q).$$

Proof. If consists essentially in the following chain of equivalences

$$Lx = y \Longleftrightarrow Lx = (I - Q)y, \ 0 = Qy$$

$$\Longleftrightarrow K_P Lx = K_P(I - Q)y, \ 0 = \Pi y$$

$$\Longleftrightarrow (I - P)x = K_P \ (I - Q)y, \ 0 = \Lambda\Pi y$$

$$\Longleftrightarrow (I - P)x = (\Lambda\Pi + K_{P,Q})y$$

where use has been made of (III.4), (III.5) and (III.2).

4. Now if P', Q' denote respectively other algebraic projectors such that the sequence

$$X \xrightarrow{P'} \operatorname{dom} L \xrightarrow{L} Z \xrightarrow{Q'} Z$$

is exact, then using (III.3) and the corresponding relation for P' we obtain

$$L(K_P - K_{P'}) = 0$$

which implies that $K_P - K_{P'}$ maps Im L into Ker L and hence

$$K_P - K_{P'} = P(K_P - K_{P'}) = P'(K_P - K_{P'})$$

from which we get at once the relations

$$PK_{P'} + P'K_P = 0 \qquad (III.6)$$

$$K_{P'} = (I-P')\,K_P \qquad (III.7)$$

B. Definition of coincidence degree for some nonlinear perturbations of Fredholm mappings in normed spaces.

1. Let now X,Z be normed real spaces, $\Omega \subset X$ a bounded open set with closure $\bar{\Omega}$, and

$$L : \text{dom } L \subset X \to Z, \qquad N : \bar{\Omega} \subset X \to Z$$

mappings such that :

(i) L is linear and Im L is closed in Z

(ii) Ker L and coker L have finite dimension and

$$\dim \text{Ker } L = \dim \text{coker } L \quad (= \text{codim Im } L)$$

(iii) $\Pi N : \bar{\Omega} \to Z$ is continuous and $\Pi\, N(\bar{\Omega})$ is bounded.

For brevity, a mapping L satisfying (i) and (ii) will be called a Fredholm mapping index zero.

2. Now it follows from (i) - (ii) and classical results of functional analysis that continuous projectors $P : X \to X$, $Q: Z \to Z$ exist such that the sequence

$$X \xrightarrow{P} \text{dom } L \xrightarrow{L} Z \xrightarrow{Q} Z$$

is exact and P(or P', P", ...) and Q (or Q', Q", ...) will always denote in the sequel continuous projectors having the same property. Moreover, the canonical surjection $\Pi: Z \to \text{coker } L$ is continuous, with the quotient topology on coker L.

Let us now assume that

(iv) $K_{P,Q}N : \overline{\Omega} \to X$ is compact (completely continuous) in $\overline{\Omega}$, i.e. continuous and such that $K_{P,Q}N(\overline{\Omega})$ is relatively compact.

3. We shall now prove the following

Proposition III.1. If (i), (ii), (iii) are satisfied and if condition (iv) holds for some couple of projectors (P,Q), then it holds for each other couple (P',Q').

Proof. We have, if $\Pi_Q = \Pi | \operatorname{Im} Q$, $\Pi_{Q'} = \Pi | \operatorname{Im} Q'$,

$$K_{P',Q'}N = K_{P'}(I-Q')N = (I-P')K_P(I-Q)N + (I-P')K_P(Q-Q')N$$

$$= (I - P')K_{P,Q}\, N + (I - P')\tilde{K}_P\, (\Pi_Q^{-1} - \Pi_{Q'}^{-1})\Pi . N$$

where we denote by $\tilde{K}_P$ the restriction of K_P to the finite-dimensional subspace $\operatorname{Im}(Q - Q')$ of Z. Thus, by a classical result of functional analysis, $\tilde{K}_P$ is necessarily continuous and then continuity of $K_{P',Q'}N$ is immediate. On the other hand, $(I-P')K_{P,Q}N(\overline{\Omega})$ is clearly relatively compact and the same for $(I-P')\, \tilde{K}_P\, (\Pi_Q^{-1} - \Pi_{Q'}^{-1})\, \Pi\, N(\overline{\Omega})$ which is a bounded set in a finite dimensional subspace of X.

Thus assumption (iv) does not depend upon the choice of P and Q and, for brevity, a mapping $N: \overline{\Omega} \to Z$ satisfying (iii) and (iv) will be said L-compact in $\overline{\Omega}$, a concept which reduces to the classical one of compactness (or complete continuity) in $\overline{\Omega}$ if X = Z and L = I.

4. We shall be interested in proving the existence of solutions for the operator equation

$$Lx = Nx, \tag{III.8}$$

a solution being an element of $\operatorname{dom} L \cap \overline{\Omega}$ verifying (III.8).

As an immediate consequence of Proposition III.1, we have the

Proposition III.2. x is a solution of (III.8) if and only if $(I-P)x = (\Lambda\Pi + K_{P,Q})\, Nx$, where Λ: coker L $\to$ ker L is any isomorphism. In other words the set of solutions of (III.8) is equal to the set of fixed points of the mapping

$$M : \bar{\Omega} \to X,$$

defined by

$$M = P + (\Lambda\Pi + K_{P,Q})N.$$

Remark : It is to be noted that, by definition, $M(\bar{\Omega}) \subset$ dom L.

We have also the following

Proposition III.3. If assumptions (i) to (iv) holds, then M is compact in Ω.

Proof. It is a trivial consequence of (iv) for the term $K_{P,Q}N$, of the fact that P has a finite-dimensional range and of (iii) and the fact that $\Lambda\Pi$ is continuous and has a finite-dimensional range for $\Lambda\Pi N$.

Therefore, $\partial\Omega$ being the boundary of Ω, if

(v) $0 \notin (L - N)(\text{dom } L \cap \partial\Omega)$

the Leray-Schauder degree

$$d[I-M, \Omega, 0]$$

is well defined and we will study to what extent it depends upon the choice of P,Q and Λ. We shall need for that a number of definitions and lemmas.

5. Let us denote $\mathcal{L}_L$ the set of isomorphisms from coker L into ker L.

Definition III.1 : $\Lambda, \Lambda' \in \mathcal{L}_L$ will be said homotopic in $\mathcal{L}_L$ if there exists a continuous mapping $\underline{\Lambda}$: coker L x [0,1] $\to$ ker L such that $\underline{\Lambda}(.,0) = \Lambda$, $\underline{\Lambda}(.,1) = \Lambda'$ and $\underline{\Lambda}(.,\lambda) \in \mathcal{L}_L$ for each $\lambda \in [0,1]$.

To be homotopic in $\mathcal{L}_L$ is an equivalence relation which gives a partition of $\mathcal{L}_L$ in homotopy classes.

<u>Proposition III.4. Λ and Λ' are homotopic in $\mathcal{L}_L$ if and only if</u> $\det(\Lambda'\Lambda^{-1}) > 0$.

<u>Proof</u>. <u>Necessity</u>. Let $\underline{\Lambda}$ be the mapping introduced in definition III.1 and $[a_1, \ldots, a_n]$, $[b_1, \ldots, b_n]$ be bases in coker L and ker L respectively. Then if $\Delta(\lambda)$ is the determinant of the matrix which represents $\Lambda(.,\lambda)$ in those bases we have $\Delta(\lambda) \neq 0$, $\lambda \in [0,1]$ and hence $\Delta(0)$ and $\Delta(1)$ have the same sign because Δ is continuous in λ. Hence, $\det(\Lambda'\Lambda^{-1}) = \frac{\Delta(1)}{\Delta(0)} > 0$.

<u>Sufficiency</u>. Let $\tilde{\Lambda}$, $\tilde{\Lambda}'$ be matrix-representations of Λ and Λ' in the respective bases $[a_1, \ldots, a_n]$, $[b_1, \ldots, b_n]$ in coker L and ker L. Then by assumption, $\det \tilde{\Lambda}$ and $\det \tilde{\Lambda}'$ have the same sign and thus belong to the same connected component of the topological group $GL(n,R)$. $GL(n,R)$ being locally arcwise connected, the corresponding component is arcwise connected and hence there exists a continuous mapping $\lambda \to \bar{\Lambda}(\lambda)$ of $[0,1]$ into $GL(n,R)$ such that $\bar{\Lambda}(0) = \tilde{\Lambda}$, $\bar{\Lambda}(1) = \tilde{\Lambda}'$. Taking for $\underline{\Lambda}(.,\lambda)$ the family of isomorphisms from coker L into ker L defined by this continuous family of matrices completes the proof.

<u>Corollary III.1. $\mathcal{L}_L$ is partitioned into two homotopy classes</u>.

Let us now chose an orientation on coker L and on ker L and let $[a_1, \ldots, a_n]$ be a basis in coker L belonging to the chosen orientation.

<u>Definition III.2</u>. Λ : coker L $\to$ ker L will be said an <u>orientation preserving</u> isomorphism if $[\Lambda a_1, \ldots, \Lambda a_n]$ belongs to the orientation chosen in ker L. If not, Λ will be said <u>orientation reversing</u>.

<u>Proposition III.5. If coker L and ker L are oriented, then</u> Λ <u>and</u> Λ' <u>are homotopic in $\mathcal{L}_L$ if and only if they are simultaneously orientation preserving or orientation reversing isomorphisms.</u>

Proof. Let $[a_1,\ldots,a_n],[b_1,\ldots,b_n]$ be respectively bases in coker L and ker L belonging to the chosen orientations. By a well-known result of linear algebra, the basis $[\Lambda a_1,\ldots, \Lambda a_n]$ in ker L will belong the orientation of $[b_1,\ldots,b_n]$ if and only if the matrix $S = (s_{ij})$ defined by

$$\Lambda a_i = \sum_{j=1}^{n} s_{ji} b_j$$

is such that det $S > 0$. Let S' be the corresponding matrix for $[\Lambda' a_1,\ldots,\Lambda' a_n]$. Then, if $G = (g_{ij})$ represents $\Lambda'\Lambda^{-1}$ in $[b_1,\ldots,b_n]$,

$$\sum_{j=1}^{n} s'_{ji} b_j = \Lambda' a_i = (\Lambda'\Lambda^{-1})\Lambda a_i = \sum_{k=1}^{n} s_{ki}(\Lambda'\Lambda^{-1})(b_k)$$

$$= \sum_{k=1}^{n} s_{ki} \sum_{j=1}^{n} g_{jk} b_j$$

which implies that

$$S' = GS.$$

The result then follows immediately,

6. Lemma III.1. *If* Y *is a vector space and* $S,S' : Y \to Y$ *two projectors such that*

$$\text{Im } S = \text{Im } S' \neq \{0\},$$

then $S'' = aS + bS'$, $a,b \in \mathbb{R}$, *is a projector having the same range if and only if*

$$a + b = 1 \qquad \text{(III.9)}$$

Proof. Necessity. If $S'' = aS + bS'$ is a projector with range equal to Im S, then

$$aS + bS' = S'' = (S'')^2 = (a+b)\,(aS + bS').$$

Thus, for each $x \neq 0$ in Im S = Im S',

$$(a + b)x = (a + b)^2\, x$$

which implies (III.9) because a + b cannot be zero.

Sufficiency. If a + b = 1, the above computation shows that $S'' = (S'')^2$. Moreover, for each $x \in$ Im S,

$$S''x = aSx + bS'x = x,$$

which shows that Im S" = Im S.

Lemma III.2. If P, P' are continuous projector onto ker L and if P" = aP + bP' with a + b = 1, then

$$K_{P''} = aK_P + bK_{P'}.$$

Proof. Using (III.7) with P" and P we get

$$K_{P''} = (I - aP - bP')\,K_P = K_P - bP'K_P$$

$$= aK_P + b(I - P')\,K_P = aK_P + bK_{P'}.$$

7. We are now ready to prove the basic

Proposition III.6. If conditions (i) to (v) hold, then $d[I - M, \Omega, 0]$ depends only upon L, N, Ω and the homotopy class of Λ in $\mathcal{L}_L$.

Proof. Let P, P' and Q, Q' be respectively continuous projectors such that

$$\text{Im } P = \text{Im } P' = \ker L,\ \ker Q = \ker Q' = \text{Im } L$$

and Λ, Λ' be two isomorphisms between coker L and ker L which belong to the same homotopy class. Let $\underline{\Lambda}$: coker L x [0,1] → ker L be the corresponding mapping in Definition III.1. It follows from Lemma III.1 that, for each $\lambda \in [0,1]$,

$$P(\lambda) = (1 - \lambda)P + \lambda P',\quad Q(\lambda) = (1-\lambda)Q + \lambda Q'$$

are respectively continuous projectors such that

$$\operatorname{Im} P(\lambda) = \ker L, \ \ker Q(\lambda) = \operatorname{Im} L$$

and moreover

$$P(0) = P, \ P(1) = P', \ Q(0) = Q, \ Q(1) = Q'.$$

Hence, by Lemma III.2,

$$K_{P(\lambda)} = (1 - \lambda)\, K_P + \lambda K_{P'}.$$

On the other hand it follows from Proposition III.2 that for each $\lambda \in [0,1]$, the fixed points of the mapping

$$\underline{M}(.,\lambda) : \bar{\Omega} \to X,$$

defined by

$$\underline{M}(.,\lambda) = P(\lambda) + \underline{\Lambda}[\Pi N(.), \lambda] + K_{P(\lambda),Q(\lambda)}N$$

coincide with the solutions of (III.8). Hence, by condition (v),

$$x \neq \underline{M}(x,\lambda) \qquad \forall\, x \in \partial\Omega, \qquad \forall\, \lambda \in [0,1].$$

Of course,

$$\underline{M}(.,0) = M$$

and

$$\underline{M}(.,1) = M' = P' + (\Lambda'\Pi + K_{P',Q'}N).$$

Let us show now that the mapping

$$(x,\lambda) \longmapsto M(x,\lambda)$$

is compact in $\bar{\Omega} \times [0,1]$. From its explicit from

$$\underline{M}(x,\lambda) = (1-\lambda)\, Px + \lambda P'x + \underline{\Lambda}[\Pi N(.),\lambda] + [(1-\lambda)K_P + \lambda K_{P'}][I - (1-\lambda)Q - \lambda Q']N$$

it is easily seen that it is continuous. To show that $\underline{M}(\bar{\Omega} \times [0,1])$ is relatively compact, the only delicate point is the last term which can be written, using (III.7),

$$(I - \lambda P')K_P(I - Q)N + \lambda(I - \lambda P')K_P(Q - Q')N.$$

The result is then easily proved by the same arguments than in Proposition III.3.

Hence, by the invariance of Leray-Schauder degree with respect to a compact homotopy, we obtain

$$d[I - M, \Omega, 0] = d[I - \underline{M}(.,0), \Omega, 0] = d[I-\underline{M}(.,1), \Omega, 0]$$
$$= d[I - M', \Omega, 0],$$

and the proof is complete.

8. Now we may explicit how $d[I - M, \Omega, 0]$ depends upon the homotopy class of Λ.

Lemma III.3. *If* $G : \ker L \to \ker L$ *is any automorphism and if*

$$M' = P + (G\Lambda\Pi + K_{P,Q})N,$$

then

$$I - M' = (I - P + GP)(I - M)$$

Proof. Compute.

Proposition III.7. *If* $\Lambda, \Lambda' \in \mathcal{L}_L$ *and if*

$$M' = P + (\Lambda'\Pi + K_{P,Q})N,$$

then

$$d[I - M', \Omega, 0] = \text{sign det } (\Lambda'\Lambda^{-1})d[I-M, \Omega, 0].$$

Proof. By the above lemma,

$$I - M' = [I - P + (\Lambda'\Lambda^{-1})P](I-M)$$

and it is easily checked that the compact perturbation of identity

$$I - P + (\Lambda'\Lambda^{-1})P$$

is an automorphism of X. Then by Leray product theorem,

$$d[I - M', \Omega, 0] = d[I - P+(\Lambda'\Lambda^{-1})P,B(1),0].d[I-M, \Omega, 0]$$

with B(1) the unit open ball with center 0 in X. On the other hand, by definition of Leray-Schauder degree,

$$d[I-P+(\Lambda'\Lambda^{-1})P, B(1), 0] = d[I-P+(\Lambda'\Lambda^{-1})|\ker L, B(1) \cap \ker L, 0]$$

$$= d[\Lambda'\Lambda^{-1}, B(1) \cap \ker L, 0] = \text{sign dét } (\Lambda'\Lambda^{-1}).$$

Corollary III.2. Under assumptions of Proposition III.6, $|d[I - M, \Omega, 0]|$ depends only upon L, N and Ω.

9. Now if we fix an orientation on ker L and coker L we may introduce the following

Definition III.3. If L, N, Ω satisfy conditions (i) to (v), the coincidence degree $d[(L,N),\Omega]$ of L and N in Ω is the integer

$$d[(L,N), \Omega] = d[I - M, \Omega, 0]$$

where, in M, Λ is an orientation preserving isomorphism.

The definition is justified by the above lemmas.

C. Remarks on the definition of coincidence degree and basic properties.

1. Let us first remark that, if $X = Z$ and $L = I$, assumptions (i)(ii) are trivially satisfied with

$$\dim \ker L = \dim \operatorname{coker} L = 0.$$

Thus $P = 0$, $Q = 0$, $K_{P,Q} = I$ and (iii) and (iv) reduce to the compactness of N on $\bar{\Omega}$. Condition (v) means in this case that N has no fixed point on $\partial\Omega$ and $M = N$. Thus

$$d[(I,N),\Omega] = d[I-N,\Omega,0],$$

and coincidence degree of I and N is nothing but Leray-Schauder degree of $I-N$.

2. The term coincidence degree has been chosen to emphasize the fact that this concept is not invariant with respect to the various possible decompositions of a mapping F into the difference of a Fredholm mapping L of index zero and a L-compact mapping N. However, it can be shown that the absolute value of the coincidence degree remains invariant. We shall not prove this fact here.

3. Now we can give the basic properties of coincidence degree.

Theorem III.1. Let conditions (i) to (v) be satisfied.

(1) (Existence theorem). If $d[(L,N),\Omega] \neq 0$, then $0 \in (L-N)(\operatorname{dom} L \cap \Omega)$.

(2) (Excision property). If $\Omega_0 \subset \Omega$ is an open set such that

$$(L - N)^{-1}(0) \subset \Omega_0,$$

then

$$d[(L,N),\Omega] = d[(L,N),\Omega_0].$$

(3) (Additivity property). If $\Omega = \Omega_1 \cup \Omega_2$, with Ω_1, Ω_2 open and such that $\Omega_1 \cap \Omega_2 = \emptyset$, then

$$d[(L,N),\Omega] = d[(L,N),\Omega_1] + d[(L,N),\Omega_2].$$

(4) (*Generalized Borsuk theorem*). *If* Ω *is symmetric with respect to* 0 *and contains it, and if* $N(-x) = -Nx$ *in* Ω, *then* $d[(L,N),\Omega]$ *is odd.*

Proof. All those results are trivial consequences of the definition of coincidence degree and of the corresponding properties of Leray-Schauder degree.

4. We also have a property of *invariance with respect to homotopies*.

Theorem III.2. *If* L *is a Fredholm mapping of index zero, if* $\tilde{N}: \bar{\Omega} \times [0,1] \to Z$, $(x,\lambda) \longmapsto \tilde{N}(x,\lambda)$ *is L-compact in* $\bar{\Omega} \times [0,1]$ *and such that, for each* $\lambda \in [0,1]$,

$$0 \notin [L - \tilde{N}(.,\lambda)](\operatorname{dom} L \cap \partial\Omega)$$

then $d[(L,N(.,\lambda)),\Omega]$ *is independent of* λ *in* $[0,1]$.

Proof. It is a trivial consequence of the invariance property of Leray-Schauder degree.

Corollary III.3. $d[(L,N),\Omega]$ *depends only upon* L,Ω *and the restricition of* N *to* $\partial\Omega$.

Proof. If N and N' are equal on $\partial\Omega$, use the homotopy

$$\tilde{N}(x,\lambda) = (1-\lambda)Nx + \lambda Nx'$$

and apply the Theorem III.1.

Corollary III.4. (*Rouché's theorem - first form*). *Let* L *be a Fredholm mapping of index zero*, N *a* L-*compact mapping in* $\bar{\Omega}$ *and suppose that* (v) *holds*. *Let* $\mu > 0$ *be defined by*

$$\mu = \inf_{x \in \partial\Omega} \|x - Px - (\Lambda\Pi + K_{P,Q})Nx\|.$$

Then, for each L-compact $N' : \bar{\Omega} \to Z$ *such that*

$$\sup_{x \in \partial\Omega} \|(\Lambda\Pi + K_{P,Q})(Nx - N'x)\| < \mu$$

one has

$$d[(L,N),\Omega] = d[(L,N'),\Omega]$$

Proof. First it is a classical result in the theory of nonlinear compact perturbations of identity that $\mu > 0$. Now if we consider the homotopy

$$\tilde{N}(x,\lambda) = (1-\lambda)Nx + \lambda N'x \ ,$$

we have, if $x \in \partial\Omega$ and $\lambda \in [0,1]$

$$\| x - Px - (\Lambda\Pi + K_{P,Q})\tilde{N}(x,\lambda) \|$$

$$\geq \| x - Px - (\Lambda\Pi + K_{P,Q})Nx \| - \lambda \| (\Lambda\Pi + K_{P,Q})(Nx - N'x) \| > 0$$

which shows that $Lx \neq \tilde{N}(x,\lambda)$ for each $x \in \partial\Omega \cap \operatorname{dom} L$ and each $\lambda \in [0,1]$. Then theorem III.2 implies the result.

5. By assuming a little more on L, it is possible to give another form to Rouché's theorem which does not involve $\Lambda\Pi$ and $K_{P,Q}$ and hence is more intrinsic. This requires the two following Lemmas.

Lemma III.4. For each $x \in \operatorname{dom} L \cap \overline{\Omega}$ we have

$$(I - M)x = (\Lambda\Pi + K_{P,Q})\ (L - N)$$

where $\Lambda\Pi + K_{P,Q}$ is an algebraic isomorphism between Z and dom L.

Proof. We have

$$(\Lambda\Pi + K_{P,Q})(L - N) = -\Lambda\Pi N + K_P L - K_{P,Q}N =$$

$$-\Lambda\Pi N + I - P - K_{P,Q}N = I - M.$$

To show that $\Lambda\Pi + K_{P,Q}$ is an algebraic isomorphism from Z onto dom L, let $y \in \operatorname{dom} L$ and consider the equation

$$(\Lambda\Pi + K_{P,Q})z = y.$$

It is equivalent to

$$\Lambda\Pi z = Py, \qquad K_{P,Q}z = (I - P)y.$$

Or, using the fact that $\Pi(I - Q)z = 0$ and that $\Lambda\Pi_Q = \Lambda\Pi|\operatorname{Im} Q$ is an isomorphism onto ker L,

$$Qz = (\Lambda\Pi_Q)^{-1}Py, \ (I-Q)z = L(I-P)y = Ly$$

and hence

$$z = [(\Lambda\Pi_Q)^{-1}P + L]y$$

which shows that

$$(\Lambda\Pi + K_{P,Q})^{-1} = (\Lambda\Pi_Q)^{-1}P + L$$

and achieves the proof.

Lemma III.5. If assumptions (i) to (v) hold for (L,N) and if $K_{P,Q}$ is continuous, then there exists $\mu > 0$ such that

$$\inf_{x\in\partial\Omega\cap\mathrm{dom}\, L} \|Lx - Nx\| \geq \mu \tag{III.10}$$

Proof. If (III.10)does not hold, there will exist a sequence (x_n) in $\partial\Omega \cap \mathrm{dom}\, L$ such that, for each $n = 1,2,\ldots,$

$$\|Lx_n - Nx_n\| \leq \frac{1}{n}.$$

Hence, using Lemma III.4 and noting that $\Lambda\Pi + K_{P,Q}: Z \to X$ is continuous, we get

$$\|x_n - Mx_n\| \leq \|\Lambda\Pi + K_{P,Q}\| \, \|Lx_n - Nx_n\| \leq \frac{k}{n}, \tag{III.11}$$

where $k \geq 0$ is the norm of $\Lambda\Pi + K_{P,Q}$. But (Mx_n) is contained in a compact set and hence there exists a subsequence $(Mx_{n'})$ which converges to some $y \in X$. But then, by (III.11), $x_{n'}$ converges to y, which therefore belongs to $\partial\Omega$ and satisfies

$$y - My = 0.$$

Then $y \in \partial\Omega \cap \mathrm{dom}\, L$ and is such that $Ly - Ny = 0$, a contradiction.

Theorem III.3. (Generalized Rouche's theorem - second form). If assumptions (i) to (v) hold for (L,N) and $K_{P,Q}$ is continuous then, for each N' L-compact in $\bar{\Omega}$ and such that

$$\sup_{x\in\partial\Omega} \|Nx - N'x\| < \mu,$$

one has

$$d[(L,N'),\Omega] = d[(L,N),\Omega]$$

Proof. Let us consider the homotopy defined by

$$\tilde{N}(x,\lambda) = (1-\lambda)Nx + \lambda N'x, \qquad \lambda \in [0,1], \quad x \in \bar{\Omega}.$$

Clearly, $\tilde{N}$ is L-compact on $\bar{\Omega} \times [0,1]$. Now, if $x \in \partial\Omega \cap \operatorname{dom} L$, $\lambda \in [0,1]$,

$$\|Lx - \tilde{N}(x,\lambda)\| = \|Lx - Nx + \lambda(Nx - N'x)\|$$

$$\geq \|Lx - Nx\| - \lambda\|Nx - N'x\| > 0$$

which implies by Theorem III.2 that

$$d[(L,N),\Omega] = d[(L,\tilde{N}(.,0)),\Omega] = d[(L,\tilde{N}(.,1)),\Omega] = d[(L,N'),\Omega].$$

6. Bibliographical notes about Chapters III.

The material of chapter III is taken from Mawhin (J. Differential Equations 12 (1972) 610-636) where it is given for mappings between locally convex topological vector spaces. See also Mawhin (Rapport n° 64, Sémin. Math. Appl. Méc. Univ. Louvain, 1973).

For short introductions to Leray-Schauder's degree theory used here see Berger and Berger ("Perspectives in Nonlinearity", Benjamin, 1968) or Rouche et Mawhin ("Equations différentielles ordinaires", tome II, ch. XI, Masson, 1973). For more complete exposition see the still so up-to-date paper of Leray-Schauder (Ann. Ec. Norm. Sup. (3) 51 (1934) 45-78) or Nagumo's one (Amer. J. Math. 73(1951) 497-511) and the books by Krasnosel'skii ("Topological Methods in the Theory of Nonlinear Integral Equations", Pergamon, 1963), Cronin ("Fixed points and topological degree in nonlinear analysis", AMS, 1964), Schwartz ("Nonlinear Functional Analysis", Gordon and Breach, 1969).

More details about section C of Chapter III can be found in Mawhin (Ann. Soc. Sci. Bruxelles 87 (1973) 51-57; rapp. n° 64, Sém. Math. Appl. Méc. Univ. Louvain, 1973). This last work can be used as a reference for section C.3 and C.4 as well as Mawhin (J. Differential Equ. 12 (1972) 610-639). Lemma III.5 Corollary III.4 and Theorem III.3 are given here for the first time.

IV. A GENERALIZED CONTINUATION THEOREM AND EXISTENCE THEOREMS FOR $Lx = Nx$

1. Let us first introduce a Brouwer degree in zero-dimensional vector spaces. If $\{0\}$ is such a space, we have only the identity mapping $I: \{0\} \to \{0\}$, $0 \longmapsto 0$ and we shall define its Brouwer degree with respect to $\{0\}$ and 0 by

$$d[I,\{0\},0] = 1$$

which clearly agrees with the usual meaning of the degree. Also we will set

$$d[I,\phi,0] = 0,$$

which completes our knowledge of degree in zero-dimensional space.

2. We will now extend, in the frame of coincidence degree theory, the well-known Leray-Schauder continuation theorem. Let $L: \operatorname{dom} L \subset X \to Z$ be a Fredholm mapping of index zero and

$$N^*: \bar{\Omega} \times [0,1] \to Z, \qquad (x,\lambda) \longmapsto N^*(x,\lambda)$$

be a L-compact mapping in $\bar{\Omega} \times [0,1]$ and let us write $N = N^*(.,1)$. Let $y \in \operatorname{Im} L$ and consider the family of equations

$$Lx = \lambda N^*(x,\lambda) + y, \qquad \lambda \in [0,1]. \qquad \text{(IV.1)}$$

Lemma IV.1. *For each $\lambda \in]0,1]$, the set of solutions of equation (IV.1) is equal to the set of solutions of equation*

$$Lx = QN^*(x,\lambda) + \lambda(I-Q)N^*(x,\lambda) + y, \qquad \text{(IV.2)}$$

and, if $\lambda = 0$, every solution of (IV.2) is a solution of (IV.1).

Proof. If $\lambda \in]0,1]$, (IV.1) is equivalent to

$$0 = QN^*(x,\lambda) , \qquad Lx = \lambda(I-Q)N^*(x,\lambda) + y ,$$

and hence to (IV.2). If $\lambda = 0$, (IV.2) is equivalent to

$$0 = QN^*(x,0), \qquad Lx = y$$

and the result is clear.

Now we can prove the basic

Theorem IV.1. (generalized continuation theorem). Let L and N^* be like above and such that the following conditions are verified :

(1) $Lx \neq \lambda N^*(x,\lambda) + y$ for every $x \in \text{dom } L \cap \partial\Omega$ and every $\lambda \in]0,1[$;

(2) $\Pi N^*(x,0) \neq 0$ for every $x \in L^{-1}\{y\} \cap \partial\Omega$;

(3) $d[\Pi N^*(.,0)|L^{-1}\{y\}, \Omega \cap L^{-1}\{y\}, 0] \neq 0$

where this last number is the Brouwer degree at $0 \in \text{coker } L$ of the continuous mapping $\Pi N^*(.,0)$ from the affine finite dimensional topological space $L^{-1}\{y\}$ into coker L. Recall that it is just the usual Brouwer degree of the mapping from R^n into itself obtained from $\Pi N^*(.,0)|L^{-1}\{y\}$ when bases have been chosen in $L^{-1}\{y\}$ and coker L and that the sign of this degree depends upon the orientations chosen on the spaces.

Then, for each $\lambda \in [0,1[$, equation (IV.1) has at least one solution in Ω and equation

$$Lx = Nx + y$$

has at least one solution in $\bar{\Omega}$.

Proof. We shall apply Theorem III.2 in the special case where

$$\tilde{N}(x,\lambda) = QN^*(x,\lambda) + \lambda(I-Q)N^*(x,\lambda) + y,$$

which clearly is L-compact. Moreover, by assumption (1) and Lemma IV.1, $Lx \neq \lambda N^*(x,\lambda) + y$ for each $x \in \partial\Omega \cap \text{dom } L$ and each $\lambda \in]0,1[$. If $\lambda = 0$, equation (IV.2) is equivalent to

$$Lx = y, \qquad QN^*(x,0) = 0$$

or

$$\Pi N^*(x,0) = 0, \quad x \in L^{-1}\{y\} \qquad \text{(IV.3)}$$

By assumption (2), no solution of (IV.3) exists which is in $\partial\Omega$ and hence

$$Lx \neq \lambda N^*(x,\lambda) + y \qquad \text{(IV.4)}$$

for each $x \in \partial\Omega \cap \text{dom } L$ and each $\lambda \in [0,1]$. Now, if there exists $x \in \partial\Omega \cap \text{dom } L$ such that

$$Lx = N^{*}(x,1) + y = Nx + y$$

the last part of the theorem is proved. If not (IV.4) now holds for each $\lambda \in [0,1]$ and each $x \in \partial\Omega \cap \text{dom } L$. Hence, by Theorem III.2, $d[L,\tilde{N}(.,\lambda)),\Omega]$ is independent of λ in $[0,1]$ and hence equal to its value at $\lambda = 0$.

But

$$d[(L,\tilde{N}(.,0)),\Omega] = d[(L,QN^{*}(.,0)+y),\Omega]$$

$$= d[I - P - \Lambda\Pi N^{*}(.,0) - K_P y,\ \Omega,\ 0].$$

If $\ker L = \{0\}$, $P = 0$, $Q = 0$, $\Pi = 0$, $K_{P,Q} = L^{-1}$ and hence

$$d[(L,\tilde{N}(.,0)),\Omega] = d[I-L^{-1}y,\Omega,0].$$

But assumption (2) in this case clearly is satisfied if and only if

$$L^{-1}\{y\} \cap \partial\Omega = \emptyset, \qquad \text{(IV.5)}$$

and, with our conventions on degree between zero-dimensional spaces, assumption (3) will be satisfied if and only if

$$\Omega \cap L^{-1}\{y\} \neq \emptyset$$

i.e. if and only if $L^{-1} y \in \Omega$, which immediately implies (IV.5).

But then

$$d[I - L^{-1}y,\Omega,0] = d[I,\Omega,L^{-1}y] = 1$$

and the theorem follows from the Theorem III.1.

If $\ker L \neq \{0\}$, then from the invariance of Leray-Schauder degree by translation and from its definition,

$$d[I - P - \Lambda\Pi N^{*}(.,0) - K_P y,\ \Omega,0]$$

$$= d[I - P - \Lambda\Pi N^{*}(.+K_P y,0),\ -K_P y + \Omega,\ 0]$$

$$= d[\{I - P - \Lambda\Pi N^{*}(.+K_P y,0)\}|\ker L,\ (-K_P y+\Omega) \cap \ker L,0]$$

$$= d[-\Lambda\Pi N^{*}(. + K_P y,0)|\ker L,\ (-K_P y+\Omega) \cap \ker L,0]$$

$$= \pm\, d[\Pi N^{*}(. + K_p y, 0)|\ker L, (-K_p y + \Omega) \cap \ker L, 0]$$

$$= \pm\, d[\Pi N^{*}(., 0)|L^{-1}\{y\}, \Omega \cap L^{-1}\{y\}, 0] \qquad (IV.6)$$

where we have used the multiplicative property of Brouwer degree, its invariance by translation and the fact the degree of a linear isomorphism is equal to ± 1. By assumption (3), the last term in (IV.6) is different of zero and the proof is complete.

3. Conditions (2) and (3) implying quotient space can seem difficult to verify in practice. On can give an equivalent form which avoids the introduction of quotient spaces.

Corollary IV.1. Suppose that condition (1) of Theorem IV.1 holds and that for some continuous projector $Q: Z \to Z$ such that $\ker Q = \operatorname{Im} L$ and some isomorphism $J : \operatorname{Im} Q \to \ker L$ one has

(2') $QN^{*}(x,0) \neq 0$ for every $x \in L^{-1}\{y\} \cap \partial\Omega$;

(3') $d[JQN^{*}(.+K_p y)|\ker L, (-K_p y+\Omega) \cap \ker L, 0] \neq 0$,

then the conclusions of Theorem IV.1 hold.

Proof. It follows directly from (IV.6) if one notes that

$$\Lambda\Pi N^{*}(.+K_p y, 0) = \Lambda\Pi Q N^{*}(. + K_p y, 0)$$

$$= (\Lambda\Pi_Q) Q N^{*}(. + K_p y, 0)$$

and that, as noted before, $\Lambda\Pi_Q$ is an isomorphism from Im Q onto ker L. Hence,

$$d[\Pi N^{*}(.,0)|L^{-1}\{y\}, \Omega \cap L^{-1}\{y\}, 0] =$$

$$\pm d[JQN^{*}(.+K_p y)|\ker L, (-K_p y+\Omega) \cap \ker L, 0]$$

for any isomorphism $J: \operatorname{Im} Q \to \ker L$.

4. Remarks. (1) If $y = 0$, (2') and (3') become respectively :

(2") $QN^{*}(x,0) \neq 0$ for every $x \in \Omega \cap \ker L$

(3") $d[JQN^{*}(.,0)|\ker L, \Omega \cap \ker L, 0] \neq 0$.

2) The usual Leray-Schauder continuation theorem corresponds to the case where L^{-1} exists and hence, by the proof of Theorem IV.1, to problems having a coincidence degree equal to one. Suppose now that Theorem IV.1 has been used to prove an existence theorem for equations

$$Lx = Nx \tag{IV.7}$$

with ker $L \neq \{0\}$ by using the homotopy

$$Lx = \lambda N^{*}(x,\lambda), \qquad \lambda \in [0,1] \tag{IV.8}$$

and that the corresponding Brouwer degree occuring in the corresponding condition (3) has absolute value strictly greater that one. In this case, if equation (IV.7) is written in the form

$$L'x = (L'-L)x + Nx \tag{IV.9}$$

with L' invertible and $(L'-L)$ L'-compact, any homotopy of the following form, with $N^{**}(x,1) = Nx$,

$$L'x = \lambda[(L'-L)x+N^{**}(x,\lambda)], \quad \lambda \in [0,1] \tag{IV.10}$$

will never work because one would get, by the invariance of the absolute value of coincidence degree for decompositions like (IV.9), and taking $\lambda = 1$ in (IV.8) and (IV.10),

$$|d[L',L'-L+N),\Omega]| = |d[(L,N),\Omega]|. \tag{IV.11}$$

But, from the proof of Theorem IV.1, the left-hand member of (IV.11) is one and the right hand member strictly greater than one, a contradiction.

5. Theorem IV.1 is a global one. One can deduce from Corollary III.2 an useful local existence theorem.

<u>Theorem IV.2</u>. <u>Suppose that</u> L <u>is a Fredholm mapping of index zero, and that</u> $\tilde{N}: \bar{\Omega} \times [0,\varepsilon_0] \to Z$ <u>is</u> L-<u>compact, with</u> $\varepsilon_0 > 0$.

If

$$Q\tilde{N}(x,0) \neq 0 \tag{IV.12}$$

<u>for each</u> $x \in \partial\Omega \cap \ker L$, <u>and if</u>

$$d[JQ\tilde{N}(.,0)|\ker L, \Omega \cap \ker L, 0] \neq 0, \qquad \text{(IV.13)}$$

<u>then there exists</u> $0 < \varepsilon_1 \leq \varepsilon_0$ <u>such that, for each</u> $\varepsilon \in [0,\varepsilon_1]$, <u>equation</u>

$$Lx = \varepsilon\tilde{N}(x,\varepsilon) \qquad \text{(IV.14)}$$

<u>has at least one solution in</u> Ω.

<u>Proof</u>. Let us consider the family of equations

$$Lx = Q\tilde{N}(x,\varepsilon) + \varepsilon(I-Q)\tilde{N}(x,\varepsilon), \quad \varepsilon \in [0,\varepsilon_0]$$

which is equivalent to (IV.14) when $\varepsilon > 0$ (the interesting case !). Like in Theorem IV.1, assumption (IV.12) is clearly equivalent to

$$Lx \neq Q\tilde{N}(x,0)$$

for every $x \in \operatorname{dom} L \cap \partial\Omega$ and hence there exists $\mu > 0$ such that

$$\inf_{x\in\partial\Omega} \|x - Px - (\Lambda\Pi+K_{P,Q})Q\tilde{N}(x,0)\| \geq \mu.$$

Now, from the L-compactness of $\tilde{N}$ there exists $0 < \varepsilon_1 \leq \varepsilon_0$ such that

$$\sup_{(x,\varepsilon)\in\bar{\Omega}\times[0,\varepsilon_1]} \|(\Lambda\Pi+K_{P,Q})\,[Q\tilde{N}(x,0)-Q\tilde{N}(x,\varepsilon)-\varepsilon(I-Q)\tilde{N}(x,\varepsilon)]\| < \mu,$$

and hence, using Corollary IV.2, we have, for each $\varepsilon \in [0,\varepsilon_1]$,

$$d[(L,Q\tilde{N}(.,\varepsilon)+\varepsilon(I-Q)\tilde{N}(.,\varepsilon)),\Omega] =$$

$$d[(L,Q\tilde{N}(.,0)),\Omega] = d[JQ\tilde{N}(.,0)|\ker L,\Omega \cap \ker L,0]$$

for some isomorphism J: $\operatorname{Im} Q \to \ker L$. Thus the result follows from (IV.13) and Theorem III.1.

6. Interesting existence theorems can also be deduced from Borsuk generalized theorem.

<u>Theorem IV.3</u>. <u>Let</u> L <u>be a Fredholm mapping of index zero and</u> $\tilde{N}$ <u>a</u> L-<u>compact mapping in</u> $\bar{\Omega} \times [0,1]$ <u>where</u> Ω <u>is a bounded open set symmetric with respect to origin and containing it. Suppose that</u>

$$\tilde{N}(-x,0) = -\tilde{N}(x,0)$$

for each $x \in \overline{\Omega}$ *and that*

$$Lx \neq \tilde{N}(x,\lambda)$$

for each $x \in \text{dom } L \cap \partial\Omega$ *and each* $\lambda \in [0,1]$. *Hence each equation*

$$Lx = N(x,\lambda),$$

with $\lambda \in [0,1]$, *has at least one solution in* Ω.

Proof. We can use Theorem III.2 to obtain

$$d[(L,\tilde{N}(.,\lambda)),\Omega] = d[(L,\tilde{N}(.,0)),\Omega], \qquad \lambda \in [0,1]$$

and the right-hand member is different from zero by Theorem III.1(4).

Theorem IV.4. *Let* L *be a Fredholm mapping of index zero and* $\tilde{N}: \overline{\Omega} \times [0,\varepsilon_0] \to Z$ *a* L-*compact mapping with* $\varepsilon_0 > 0$ *and* Ω *open, bounded, symmetric with respect to the origin and containing it. Then if for each* $x \in \overline{\Omega}$,

$$\tilde{N}(-x,0) = -\tilde{N}(x,0)$$

and if, for each $x \in \partial\Omega \cap \text{dom } L$

$$Lx \neq \tilde{N}(x,0)$$

then there exists $0 < \varepsilon_1 \leq \varepsilon_0$ *such that for every* $\varepsilon \in [0,\varepsilon_1]$, *equation*

$$Lx = \tilde{N}(x,\varepsilon)$$

has at least one solution in Ω.

Proof. It follows from Theorem III.1(4) and Corollary III.2 and details are left to the reader.

7. We can give now another consequence of generalized Borsuk theorem, of global nature.

Theorem IV.5. (*Generalized Krasnosel'skii theorem*). *Let* L *be a Fredholm mapping of index zero,* N *a* L-*compact mapping in* $\overline{\Omega}$ *with* Ω *open, bounded, symmetric with respect to the origin and containing it*. *Then if*

$$(L-N)(x) \neq \mu(L-N)(-x)$$

<u>for every</u> $\mu \in [0,1]$ <u>and every</u> $x \in \operatorname{dom} L \cap \partial\Omega$, <u>equation</u>

$$Lx = Nx$$

<u>has at least one solution in</u> Ω.

<u>Proof</u>. Let us define $\tilde{N}:\bar{\Omega} \times [0,1] \to Z$ by

$$\tilde{N}(x,\lambda) = (1 + \lambda)^{-1}[N(x) - \lambda N(-x)].$$

We have

$$\tilde{N}(x,0) = Nx$$

and

$$\tilde{N}(x,1) = (\tfrac{1}{2})[N(x) - N(-x)]$$

is odd. Now conditions of Theorem III.2 are satisfied because if there exists $x \in \partial\Omega \cap \operatorname{dom} L$ and $\lambda \in [0,1]$ such that

$$Lx = \tilde{N}(x,\lambda),$$

then

$$(1 + \lambda)Lx = Nx - \lambda N(-x)$$

i.e.

$$(L-N)(x) = \lambda(L-N)(-x)$$

a contradiction. Thus,

$$d[(L,N),\Omega] = d[(L,\tilde{N}(.,0)),\Omega] = d[(L,\tilde{N}(.,1)),\Omega] \neq 0$$

and the proof is complete.

8. It will be noted that Theorems IV.1, IV.3 and IV.5 are special cases of the following general result which follows directly from be basic properties of coincidence degree. The interest of Theorems IV.1 and IV.3 is to involve conditions much more easy to be verified in applications.

<u>Theorem IV.6.</u> <u>Let L be a Fredholm mapping of index zero and</u> $\tilde{N}: \bar{\Omega} \times [0,1] \to Z$ <u>a L-compact mapping, with</u> Ω <u>open and bounded</u>. <u>If, for each</u>

$\lambda \in [0,1]$ and each $x \in \partial\Omega \cap \text{dom } L$,

$$Lx \neq \tilde{N}(x,\lambda)$$

and if

$$d[(L,\tilde{N}(.,0)), \Omega] \neq 0,$$

then, for each $\lambda \in [0,1]$, equation

$$Lx = \tilde{N}(x,\lambda)$$

has at least one solution in Ω.

In the same way, Theorems IV.2 and IV.4 are special but more easily used cases of the following general consequence of basic properties of coincidence degree.

Theorem IV.7. Let L be a Fredholm mapping of index zero and $\tilde{N}: \overline{\Omega} \times [0,\varepsilon_0] \to Z$ a L-compact mapping with $\varepsilon_0 > 0$ and Ω open and bounded. If, for each $x \in \partial\Omega \cap \text{dom } L$

$$Lx \neq \tilde{N}(x,0)$$

and if

$$d[(L,\tilde{N}(.,0)), \Omega] \neq 0,$$

then there exists $\varepsilon_1 \in]0,\varepsilon_0]$ such that, for each $\varepsilon \in [0,\varepsilon_1]$, equation

$$Lx = \tilde{N}(x,\varepsilon)$$

has at least one solution in Ω.

9. Bibliographical notes about Chapter IV.

The classical Leray-Schauder continuation theorem appears in Ann. Ec. Norm. Sup. (3) 51(1934) 45-78. Theorem IV.1 seems to have been given the first time, in the case of periodic solutions of ordinary differential equations, by Mawhin (Bull. Soc. R. Sci. Liège 38 (1969) 308-398) where it is proved using Cesari's method discussed in Chapter II. More direct proofs are given by Strygin (Math. Notes Acad. Sci. USSR 8 (1970) 600-602) and by Mawhin (Bull. Ac. R. Belgique, Cl. Sci. (5) 55(1969) 934-947 ; Equa-Diff 70, Marseille, 1970; J. Differential Equations 10 (1971) 240-261) for periodic solutions of ordinary and functional differential equations, Mawhin's proof being in the spirit of the one given here. The case of operator equations in Banach spaces is given in Mawhin, Rapp. Sém. Math. Appl. Méc. Univ. Louvain n° 39, 1971 and integrated in the frame of coincidence degree theory, for locally convex spaces, in Mawhin, J. Differential Equ. 12 (1972) 610-636. For the local theorem for periodic solutions, see Mawhin (Bull. Soc. R. Sci. Liège 38 (1969) 308-398) and Strygin (op. cit.).

Theorem IV.3, in the case of periodic solutions, is due to Gussefeldt (Math. Nachr. 36 (1968) 231-233). A simpler proof and extensions to more general equations are given by Mawhin in the papers quoted above. Theorem IV.5 is given in Mawhin (Rapp. Sém. Math. Appl. Méc. Univ. Louvain n° 64, 1973) and generalizes a result of Krasnosel'skii (Amer. Math. Soc. Transl. (2) 10 (1958) 345-409) corresponding to $X = Z$ and $L = I$.

V. TWO-POINT BOUNDARY VALUE PROBLEMS : NONLINEARITIES WITHOUT SPECIAL STRUCTURE

We consider various boundary value problems of the form

$$x' = f(t,x)$$
$$(x(a),x(b)) \in S \qquad (V.0)$$

where $f : [a,b] \times R^n \times R^n$ and $S \subset R^n \times R^n$. We will be particularly interested in the problems consisting of

$$x'' = f(t,x,x') \qquad (V.1)$$

together with

$$x(a) = x(b) = 0 \qquad \text{(Picard Problem)} \qquad (V.2)$$
$$x(0) = x(T),\ x'(0) = x'(T) \qquad \text{(Periodic Problem)} \qquad (V.3)$$

where $f : [a,b] \times R^n \times R^n \to R^n$ or $f : [0,T] \times R^n \times R^n \to R^n$.
By standard devices these latter problems may be written as special cases of (V.0). We will assume throughout that f is continuous on its domain.

The quest is for an existence theory for such problems which :

a) Employs sufficient conditions for existence of the greatest possible generality.
b) Employs hypotheses which may be readily verified for a given problem.
c) Provides an accompanying mechanism for approximating the solutions.

In Part V we present an existence theory which partially meets objectives a) and b). Consistent with the pursuit of a) we concentrate on methods and results which apply even when the nonlinear function f does not display such special structure as : monotonicity, quasilinearity, sublinearity, differentiability, or "Lipschitzicity". We thus take a very general approach whose theoretical framework is provided by the continuation theorem developed in Part IV (see Theorem IV.1 and Corollary IV.1).
A central feature of this approach is the topological-geometric description of certain sets G in n-space which contain solution trajectories, or more generally, the description of sets Ω in an underlying function space which

contain solutions. We use this approach to unify recent results of several authors. Specific bibliographical information concerning these results will appear at the end of Part V.

In Part VI, we use projection methods to obtain a theory of approximation for the solutions whose existence is established in Part V. The material presented here partially meets objective c) and is related to the book of Krasnosel'skii, Vainikko, Zabreiko, Rutitskii, (Approximate Solution of Operator, Equations, 1972, Noordhoff) and the recent theses of Strasberg (La Recherche de Solutions Périodique d'Equations Différentielles Non Linéaires, Univ. Libre de Bruxelles, 1975) and Chen (Constructive Methods for Nonlinear Boundary Value Problems, Colorado State University, 1974).

We begin by developing an existence theory for problems (V.1) - (V.2) and (V.1) - (V.3). In section 7 we comment on extensions of the theory to more general subclasses of (V.0).

In Section 1 we convert problems (V.1) - (V.2) and (V.1) - (V.3) to equations of the form Lx = Nx where L and N are linear and nonlinear transformations on appropriate function spaces. The continuation theorem, which gives sufficient conditions for existence of a solution to Lx = Nx in a set Ω and which provides the theoretical framework of our approach, is then briefly described. In Section 2 we provide the basis for a technique for finding appropriate *a priori* bound sets Ω. The techniques are then illustrated for the scalar cases of (V.1) - (V.2) and (V.1) - (V.3) in Sections 3, 4 and 5 and for systems in Section 6. Finally, in Section 8 we consider some specific examples. Specific bibliographical information appears at the end.

If we define

$$\begin{aligned} X &= C^1[a,b] \cap \{x : x(a) = x(b) = 0\} \\ Z &= C\,[a,b] \qquad\qquad (V.4) \\ \text{dom } L &= X \cap C^2[a,b] \\ L &: \text{dom } L \to Z, \quad x \mapsto x'' \\ N &: X \to Z, \quad x \mapsto f(\cdot,x(\cdot),x'(\cdot)), \end{aligned}$$

then the Picard problem may be written as

$$L\,x = N\,x. \qquad (V.5)$$

If we define

$$\begin{aligned} X &= C^1[0,T] \cap \{x : x(0) = x(T),\ x'(0) = x'(T)\} \\ Z &= C\,[0,T] \\ \text{dom } L &= X \cap C^2[0,T] \qquad\qquad (V.6) \end{aligned}$$

$$L : \operatorname{dom} L \to Z, \quad x \mapsto x''$$
$$N : X \to Z, \quad x \mapsto f(\cdot, x(\cdot), x'(\cdot)),$$

then the periodic problem may also be written as $Lx = Nx$. (The spaces C, C^1, C^2 are the usual Banach spaces with the usual norms.)

The two problems thus formulated have obvious similarity in structure - but also a striking difference in structure. Namely, for the Picard problem, L^{-1} exists and for the periodic problem L^{-1} does not exist. The existence of L^{-1} presents the opportunity to write (V.5) as

$$(I - L^{-1} N)x = 0.$$

If $L^{-1} N$ is compact (and we will see that it is) this enables the application of results in nonlinear functional analysis concerning compact perturbations of identity - particularly the Schauder continuation theorem.

In order to apply such techniques to the periodic problem one may replace (V.1) by an equivalent equation

$$x'' + Ax' + Bx = f(t,x,x') + Ax' + Bx$$

in such a way that $L : x \mapsto x'' + Ax' + Bx$ is invertible, but this often imposes "unnatural" structure on the nonlinear mapping N in the revised problem. We will employ here a continuation theorem which accomodates certain cases where L^{-1} does not exist - including the periodic problem considered here - without altering the structure of the nonlinear mapping. For convenience of reference we repeat here some definitions and a simplified version of Corollary IV.1.

Definitions and Notation. Let X, Z be normed vector spaces, $L : \operatorname{dom} L \subset X \to Z$ a linear mapping, and $N : X \to Z$ a continuous mapping. The mapping L will be called a *Fredholm mapping of index zero* if

(a) $\dim \operatorname{Ker} L = \operatorname{codim} \operatorname{Im} L < +\infty$

(b) $\operatorname{Im} L$ is closed in Z.

If L is a Fredholm mapping of index 0 there exist continuous projectors $P : X \to X$ and $Q : Z \to Z$ such that

$$\operatorname{Im} P = \operatorname{Ker} L$$

$$\text{Im } L = \text{Ker } Q = \text{Im}(I-Q).$$

It follows that $L|\text{dom } L \cap \text{Ker } P : (I-P)X \to \text{Im } L$ is invertible. We denote the inverse of that map by K_P. If Ω is an open bounded subset of X, the mapping N will be called *L-compact on* $\bar{\Omega}$ if $QN(\bar{\Omega})$ is bounded and $K_P(I-Q)N : \bar{\Omega} \to X$ is compact. Since Im Q is isomorphic to Ker L there exist isomorphisms $J : \text{Im } Q \to \text{Ker } L$.

Continuation Theorem *Let* L *be a Fredholm mapping of index* 0 *and let* N *be* *L-compact on* $\bar{\Omega}$. *Suppose*

a) *For each* $\lambda \in (0,1)$, *every solution* x *of*

$$Lx = \lambda Nx.$$

is such that $x \notin \partial\Omega$.

b) $QNx \neq 0$ *for each* $x \in \text{Ker } L \cap \partial\Omega$ *and*

$$d[n, \Omega \cap \text{Ker } L, 0] \neq 0$$

where $n = JQN : \text{Ker } L \to \text{Ker } L$.

Then the equation $Lx = Nx$ *has at least one solution in* $\text{dom } L \cap \bar{\Omega}$.

Remarks.

1) We will assume throughout that f is continuous on its domain. It can then be shown that N is continuous by standard arguments.

2) For the *Picard problem* under formulation (V.4), Ker L = {0} and Im L = Z. Thus it is immediate the L is Fredholm of index 0. It is well known that

$$K_P(I-Q)Nx = L^{-1}N\,x = \int_a^b G(s,t)f(s,x(s),x'(s))ds$$

where

$$-G(s,t) = \begin{cases} (b-a)^{-1}\,(b-t)(s-a), & a \leq s \leq t \leq b \\ (b-a)^{-1}\,(t-a)(b-s), & a \leq t \leq s \leq b. \end{cases}$$

From this representation it is easy to show using the Ascoli-Arzela theorem that $L^{-1}N(\bar{\Omega})$ is compact for any bounded set Ω. In this case $Q \equiv 0$, thus $QN(\bar{\Omega})$ is bounded and N is L-compact. The hypothesis (b) is trivially satisfied if we take $0 \in \Omega$ and define $d[I, \{0\}, 0] = 1$. Thus the Continuatic Theorem in this case reduces to the Leray-Schauder continuation theorem.

3) For the *periodic problem* under the formulation (V.6) we have

$$\text{Ker } L = \{x : x(t) \equiv c,\ c \in R^n\}.$$

and

$$\text{Im } L = \{z : x'' = z(t),\ x(0) = x(T),\ x'(0) = x'(T) \quad \text{for some} \quad x \in C^2[0,T]\}.$$

But the general solution to $x'' = z(t)$ is

$$x(t) = ct + d + \int_0^t \int_0^s z(\tau)d\tau$$

and it is easily seen that this equation has periodic solutions if and only if

$$\int_0^T z(\tau)d\tau = 0.$$

Thus

$$\text{Im } L = \{z : \int_0^T z(\tau)d\tau = 0\} \cap Z,$$

and we may take as the projector

$$Q : Z \to Z, \quad z \mapsto \frac{1}{T} \int_0^T z(\tau)d\tau.$$

We have

$$\text{codim Im } L = \dim \text{Im } Q = n = \dim \text{Ker } L.$$

Moreover, Im L is a closed subspace of Z. Thus L is a Fredholm mapping of index 0.

For the projector P we may take

$$P : X \to X, \quad x \mapsto x(0).$$

It may easily be shown that

$$K_{P_z} = \int_0^T G(s,t)\, z(s)ds$$

where

$$G(x,t) = \begin{cases} -\frac{s}{T}[T-t], & 0 \leqslant s \leqslant t \\ -\frac{t}{T}[T-t], & t \leqslant s \leqslant T. \end{cases}$$

Again using the Ascoli-Arzela theorem it is not difficult to show that $\overline{K_p(I-Q)N(\bar{\Omega})}$ is compact for any bounded Ω. Moreover, $QN(\bar{\Omega})$ is clearly bounded. Thus N is L-compact.

2. In this section we begin the search for appropriate open, bounded subsets Ω for the application of the continuation theorem.

Definition V.1. If $G \subset [a,b] \times R^n \times R^n$ is open in the relative topology on $[a,b] \times R^n \times R^n$ and bounded, we will call G a *bound set* relative to (V.1) if for any $(t_0, x_0, y_0) \in \partial G$ with $t_0 \in (a,b)$ there is a function $V(t,x,y) = V(t_0, x_0, y_0 ; t,x,y)$ such that :

i) $V \in C^1([a,b] \times R^n \times R^n)$
ii) $G \subset \{(t,x,y) : V(t,x,y) < 0\}$
iii) $V(t_0, x_0, y_0) = 0$
iv) If grad V denotes the gradient of V at (t_0, x_0, y_0), then

$$\text{grad } V \cdot \begin{bmatrix} 1 \\ y_0 \\ f(t_0, x_0, y_0) \end{bmatrix} \neq 0. \qquad (V.7)$$

Theorem V.2. *Let* G *be a bound set relative* to (V.1). *If* $x(t)$ *is a solution to* (V.1) *on* $[a,b]$ *with* $(a, x(a), x'(a)), (b, x(b), x'(b)) \in G$, *then* $x \notin \partial\Omega$ *where*

$$\Omega \equiv \{x : x \in C^1[a,b] \text{ and } (t,x(t),x'(t)) \in G \text{ for } t \in [a,b]\}.$$

Proof. Suppose $x \in \partial\Omega$. Then $(t_0, x(t_0), x'(t_0)) \in \partial G$ for some $t_0 \in (a,b)$. Let $u(t) = V(t,x(t),x'(t))$ where V is as in Definition V.1. It follows from definition VI that $u(t) \in C^1[a,b]$ and has a maximum at t_0 ; and thus, $u'(t_0) = 0$. But

$$u'(t_0) = \text{grad } V \cdot \begin{bmatrix} 1 \\ x'(t_0) \\ x''(t_0) \end{bmatrix} = \text{grad } V \cdot \begin{bmatrix} 1 \\ x'(t_0) \\ f(t_0,x(t_0),x'(t_0)) \end{bmatrix} \neq 0.$$

by iv).

Remarks.

1) If G is convex, then for each $(t_0, x_0, y_0) \in \partial G$ there is a linear function which satisfies i), ii), and iii) in Definition V.1.

2) If $G = \{(t,x,y) : \tilde{V}(t,x,y) < 0\}$ for some $\tilde{V} \in C^1([a,b] \times R^n \times R^n)$ then if we take $V = \tilde{V}$ for each $(t_0, x_0, y_0) \in \partial G$ i), ii), and iii) are satisfied.

3) The conditions of Definition V.1 have an easy geometric interpretation. Condition iv) says that no trajectory passing through (t_0, x_0, y_0) may be tangent to $V(t,x,y) = 0$ at that point. Thus by i), ii), and iii) no solution trajectory lying in $\bar{G}$ can "touch" ∂G at (t_0, x_0, y_0).

4) The special case $G = G_0 \times [a,b]$ may often be treated with choices of V which are independent of t. In this case the Definition V.1 may be restated without direct reference to the t variable.

The conditions of Definition V.1, though quite general, do not fully exploit the second order nature of equation (V.1).

Definition V.2. If $G_1 \subset [a,b] \times R^n$ is open in the relative topology on $[a,b] \times R^n$ and bounded, we will call G_1 a *curvature bound set* (CBS) relative to (V.1) if for any $(t_0, x_0) \in \partial G$ with $t_0 \in (a,b)$ there is a function $V_1(t,x) = V_1(t_0, x_0; t,x)$ such that

i) $V_1 \in C^2([a,b] \times R^n)$

ii) $G_1 \subset \{(t,x) ; V_1(t,x) < 0\}$

iii) $V_1(t_0, x_0) = 0$

iv) If H denotes the Hessian matrix of V_1 at (t_0, x_0), and grad V_1 denotes the gradient at (t_0, x_0),

$$H \begin{bmatrix} 1 \\ y \end{bmatrix} \cdot \begin{bmatrix} 1 \\ y \end{bmatrix} + \text{grad } V_1 \cdot \begin{bmatrix} \\ f(t_0, x_0, y) \end{bmatrix} > 0 \qquad \text{(V.8)}$$

for all y such that

$$\text{grad } V_1 \cdot \begin{bmatrix} 1 \\ y \end{bmatrix} = 0.$$

Theorem V.4. *Let* G_1 *be a curvature bound set relative to* (V.1). *If* $x(t)$ *is a solution to* (V.1) *on* $[a,b]$ *with* $(a, x(a)), (b, x(b)) \in G_1$ *then* $x \notin \partial \Gamma_1$ *where*

$$\Gamma_1 \equiv \{x : x \in C^2[a,b] \text{ and } (t, x(t)) \in G_1 \text{ for } t \in [a,b]\}$$

Proof. Suppose $x \in \partial\Gamma_1$. Then $(t, x(t)) \in \bar{G}_1$ for $t \in [a,b]$ and $(t_0, x(t_0)) \in \partial G_1$ for some t_0. Since $(a, x(a)), (b, x(b)) \in G_1$, $t_0 \in (a,b)$. Let $V_1(t,x)$ be the function associated with (t_0, x_0) in the definition of a curvature bound set. We have from i), ii), and iii) that $u(t) = V_1(t, x(t)) \leq 0$ on $[a,b]$, $u \in C^2[a,b]$, and $u(t)$ attains an interior relative maximum at t_0. Thus

$$u'(t_0) = \text{grad } V_1 \cdot \begin{bmatrix} 1 \\ x'(t_0) \end{bmatrix} = 0,$$

and

$$u''(t_0) = H \begin{bmatrix} 1 \\ x'(t_0) \end{bmatrix} \cdot \begin{bmatrix} 1 \\ x'(t_0) \end{bmatrix} + \text{grad } V_1 \cdot \begin{bmatrix} 0 \\ x''(t_0) \end{bmatrix} \leqslant 0.$$

If we let $y = x'(t_0)$ and note that $x''(t_0) = f(t_0, x(t_0), x'(t_0))$ we have a contradiction to (V.8).

Remarks.

1) Similar remarks to 1), 2), and 4) following Theorem V.2 apply here.

2) There is a geometric interpretation of Definition V.3 in terms of curvatures. Suppose $C = (t, x(t))$ is a solution trajectory in $\bar{G}_1$ which "touches" ∂G_1 at (t_0, x_0).

By i), ii), and iii), $\text{grad } V_1 \cdot \begin{bmatrix} 1 \\ x'(t_0) \end{bmatrix} = 0$. Let

$$T = \frac{1}{\sqrt{1 + \|x'(t_0)\|^2}} \begin{bmatrix} 1 \\ x'(t_0) \end{bmatrix}$$

$$N = \frac{\text{grad } V_1}{\|\text{grad } V_1\|}$$

$$C_1 = \left\{ \begin{matrix} \text{projection of C onto} \\ \text{the plane of N and T} \end{matrix} \right\} = (d(t) \cdot T)\, T + (d(t) \cdot N)\, N + (t_0, x_0)$$

$$C_2 = \begin{pmatrix} \text{intersection of } V_1(t,x) = 0 \\ \text{with the plane of } N \text{ and } T \end{pmatrix} = \alpha\, T + \beta(\alpha) N + (t_0,\ x_0)$$

where

$$d(t) = (t-t_0,\ x(t)-x(t_0))$$

and

$$V_1(\alpha\, T + \beta\, N + (t_0,\ x_0)) = 0.$$

Then

$$\text{curvature of } C_1 = \frac{-\operatorname{grad} V_1 \cdot \begin{bmatrix} 0 \\ x''(t_0) \end{bmatrix}}{\|\operatorname{grad} V_1\| \ (1+\|x'(t_0)\|^2)}$$

$$\text{curvature of } C_2 = -\ \beta''(0) = \frac{H \begin{bmatrix} 1 \\ x'(t_0) \end{bmatrix} \cdot \begin{bmatrix} 1 \\ x'(t_0) \end{bmatrix}}{\|\operatorname{grad} V_1\| \ (1+\|x'(t_0)\|^2)}.$$

Then iv) leads to a geometric contradiction.

3) Note that in general Γ_1 is not a bounded subset of $C^1[a,b]$.

As a bounded subset of $\Gamma_1 \subset C^1[a,b]$ we introduce the following:

Definition V.5. Let G_1 be a curvature bound set relative to (V.1). If $G \subset G_1 \times R^n \subset [a,b] \times R^n \times R^n$ is open in the relative topology on $[a,b] \times R^n \times R^n$ and bounded, we will call G a *Nagumo-set* relative to (V.1) if for any $(t_0,\ x_0,\ y_0) \in \partial G$ with $t_0 \in (a,b)$ and $(t_0,\ x_0) \notin \partial G_1$ there exists a $V(t_0, x_0, y_0\ ;\ t,x,y)$ satisfying the conditions of Definition (V.1).

Theorem V.6. *Let* G *be a Nagumo-set relative to* (V.1). *If* $x(t)$ *is a solution to* (V.1) *on* $[a,b]$ *with* $(a,\ x(a),\ x'(a))$, $(b,\ x(b),\ x'(b)) \in G$, *then* $x \notin \partial\Omega$ *where*

$$\Omega = \{x : x \in C^1[a,b],\ (t,x(t),\ x'(t)) \in G \text{ for } t \in [a,b]\}.$$

Proof. Suppose $x \in \partial\Omega$ and x is a solution to (V.1) on [a,b]. Then $(t,x(t),x'(t)) \in \bar{G}$ on [a,b] and $(t_0,\ x(t_0),\ x'(t_0)) \in \partial G$ for some t_0.

Since $(a,x(a),x'(a))$, $(b,x(b),x'(b)) \in G$, $t_0 \in (a,b)$. We have $(t,x(t)) \in \bar{G}_1$ for $t \in [a,b]$; thus, if $(t_0, x(t_0)) \in \partial G_1$, $x \in \partial \Gamma_1$ which contradicts. Theorem V.4. If $(t_0, x(t_0)) \notin \partial G_1$ we may apply the argument in the proof of Theorem V.2 to obtain a contradiction.

Remarks.

1) Note that in Definition V.5 and Theorem V.6 it isn't necessary that G_1 be a curvature bound set, only that no solution to (V.1) lie on $\partial \Gamma_1$

2) Theorems V.2 and V.6 are useful for determining "candidates" for sets Ω in applying the Continuation Theorem. However, given such a Ω we are left with several technical problems :

i) We must show that solutions $x(t)$ to our boundary value problem satisfy $(a,x(a),x'(a))$, $(b,x(b),x'(b)) \in G$.

ii) Hypothesis a) of the Continuation Theorem requires that $x \notin \partial\Omega$ for *all* $\lambda \in (0,1)$.

iii) Hypothesis b) in the Continuation Theorem must be satisfied.

3. We consider

$$x'' = f(t,x,x') \tag{V.9}$$
$$x(0) = x(T),\ x'(0) = x'(T) \tag{V.10}$$

where $f : [0,T] \times R \times R \to R$ is continuous. We will have need to refer to

$$x'' = \lambda f(t,x,x'). \tag{V.9.λ}$$

Theorem V.7. *Suppose*

a) $f(t,R,0) > 0$ and $f(t,-R,0) < 0$ for some $R > 0$.
b) There exists a positive function $\psi \in C^1[0,+\infty)$ such that

$$\int_0^\infty \rho\, d\rho \,/\, \psi(\rho) = +\infty$$

and

$$|f(t,x,x')| < \psi\,(|x'|)$$

for $|x| \leq R$.

Then (V.9) - (V.10) has at least one solution $x(t)$ satisfying $|x(t)| \leq R$.

Proof. We apply the Continuation Theorem using the formulation (V.6). Note that $Lx = \lambda Nx$ is equivalent to (V.9.λ) - (V.10).

a) Construction of Ω. Let

$$G_1 = \{(t,x) : |x| < R,\ t \in [0,T]\}.$$

We show that G_1 is a curvature bound set relative to (V.9.λ) for $\lambda \in (0,1)$. For $(t_0,x_0) \in \partial G_1$, let

$$V_1(t,x) \equiv x^2 - R^2.$$

Conditions (i) - (iii) of Definition V.3 are clearly satisfied. Moreover

$$H\begin{bmatrix}1\\y\end{bmatrix}\cdot\begin{bmatrix}1\\y\end{bmatrix} + \text{grad } V_1 \cdot \begin{bmatrix}0\\ \lambda f(t_0,x_0,y)\end{bmatrix} = \begin{bmatrix}0&0\\0&2\end{bmatrix}\begin{bmatrix}1\\y\end{bmatrix}\cdot\begin{bmatrix}1\\y\end{bmatrix} + \begin{bmatrix}0\\2x_0\end{bmatrix}\cdot\begin{bmatrix}0\\ \lambda f(t_0,x_0,y_0\end{bmatrix}$$

$$= 2y^2 + 2x_0\ \lambda f(t_0,x_0,y).$$

But if

$$\text{grad } V_1 \cdot \begin{bmatrix} 1 \\ y \end{bmatrix} = 2x_0y = 0$$

then $y = 0$. Since $|x_0| = R$ and $\pm Rf(t, \pm R, 0) > 0$, condition (iv) is satisfied. Thus G_1 is a CBS relative to (V.9.λ) for $\lambda \in (0,1)$.

Let $h(x)$ be the unique solution to

$$\frac{dy}{dx} = -\frac{\psi(y)}{y} .$$

Then
$$y(-R) = N.$$

$$\int_{h(x)}^{N} \frac{\rho d\rho}{\psi(\rho)} = x + R.$$

condition (b) assures that if N is chosen sufficiently large $h(x)$ is defined on $[-R,R]$ and $h(x) > 0$ on $[-R,R]$. Let

$$G = \{(t,x,y) : |x| < R, |y| < h(x)\}.$$

Suppose $(t_0, x_0, y_0) \in \partial G$ and $(t_0, x_0) \notin \partial G_1$. Then $|x_0| < R$ and $|y_0| = h(t_0)$. Define

$$V(t,x,y) = y^2 - (h(x))^2 .$$

Then (i) - (iii) of Definition I.1 are clearly satisfied. We have

$$\text{grad } V \cdot \begin{bmatrix} 1 \\ y_0 \\ \lambda f(t_0,x_0,y_0) \end{bmatrix} = \begin{bmatrix} 0 \\ -2h(x_0)h'(x_0) \\ 2y_0 \end{bmatrix} \cdot \begin{bmatrix} 1 \\ y_0 \\ \lambda f(t_0,x_0,y_0) \end{bmatrix} =$$

$$= -2h(x_0)h'(x_0)y_0 + 2\lambda y_0 f(t_0,x_0,y_0) = 2y_0[|y_0|\frac{\psi(|y_0|)}{|y_0|} + \lambda f(t_0,x_0,y_0)].$$

By hypothesis (b) the latter quantity is nonzero for $\lambda \in (0,1)$. Thus G is a Nagumo set relative to (I.9.λ) for $\lambda \in (0,1)$. Define

$$\Omega \equiv \{x : x \in C^1[0,T], (t,x(t), x'(t)) \in G \text{ for } t \in [0,T]\}.$$

Suppose $x(t)$ is a solution to (I.9.λ) - (I.10). Suppose $x(t) \in \partial\Omega$. By Theorem I.6, $(0,x(0), x'(0)), (T,x(T), x'(T)) \in \partial G$. Suppose $x(0) = x(T) = R$. since $|x(t)| \leq R$, $x'(0) = x'(T)$ implies that $x''(0) \leq 0$.

But we must have

$$x''(0) = \lambda f(0,x(0),x'(0)) = \lambda f(0,R,0) > 0.$$

But this is a contradiction. Similarly $x(0) \neq -R$.

Thus we must have $|x'(0)| = h(x(0))$ and $|x'(T)| = h(x(T))$. Consider

$$\sigma(t) = (x'(t))^2 - (h(x(t)))^2.$$

Since $\sigma(t) \leqslant 0$, $\sigma(0) = 0$, and $\sigma(T) = 0$ we must have $\sigma'(0) \leqslant 0$, and $\sigma'(T) \geqslant 0$. But

$$\sigma'(t) = 2x'(t)\, x''(t) - 2h(x(t))\, h'(x(t))\, x'(t)$$
$$\sigma'(0) = 2x'(0)\, [\lambda f(0,x(0),x'(0)) + \psi(|x'(0)|)]$$
$$\sigma'(T) = 2x'(T)\, [\lambda f(T,x(T),x'(T)) + \psi(|x'(T)|)].$$

The two expressions on the right are nonzero and of the same sign. Thus, we again have a contradiction, and we conclude $x \notin \Omega$.

b) Behavior on the Kernel of L. We have

$$\mathrm{Ker}\, L \cap \Omega = \{x : x = c,\ |c| < R\}$$
$$\mathrm{Ker}\, L \cap \partial\Omega = \{x = R,\ x = -R\}$$
$$QN(c) = \frac{1}{T} \int_0^T f(t,c,0)\,dt.$$

For the isomorphism $J : \mathrm{Im}\, Q \to \mathrm{Ker}\, L$ we may take the natural identification Then

$$QN(R) > 0,\ QN(-R) < 0$$
$$d[QN,\ \Omega \cap \mathrm{Ker}\, L,\ 0] \neq 0.$$

Remarks.

1) In the proof we used only

$$f(t,x,y) > -\psi(|y|),\ |x| \leqslant R.$$

We could have used alternatively

$$f(t,x,y) < \psi(|y|),\ |x| \leqslant R$$

$$\begin{cases} f(t,x,y) < -\psi(|y|), & y > 0, \\ f(t,x,y) < \psi(|y|), & y < 0, \end{cases} \qquad |x| \leq R$$

$$\begin{cases} f(t,x,y) < \psi(|y|), & y > 0, \\ f(t,x,y) > -\psi(|y|), & y < 0, \end{cases} \qquad |x| \leq R$$

$$xf(t,x,y) > -|x|\ \psi(|y|), \quad 0 < |x| \leq R.$$

2) In the proof we used only

$$\int_0^\infty \frac{\rho d\rho}{\psi(\rho)} > 2R.$$

Note that if $\psi(\rho) = A\rho^2 + B$, the desired divergence occurs.

3) It isn't essential that $\psi(\rho)$ be differentiable. If $\psi(\rho)$ is continuous we may argue using appropriate maximal and minimal solutions.

Examples.

1) $x'' = \pm (x')^k + x^{2m+1} + f(t),\ k \neq 0.$

This equation has a periodic solution by Theorem V.7 and Remark 1.

2) $x'' = (x')^3 \sin x' + g(t,x).$

This equation doesn't satisfy Remark 1.

Theorem V.8. Suppose

a) There exist $\alpha(t),\ \beta(t) \in C^2[0,T]$ such that

$$\alpha(t) < \beta(t),$$
$$\alpha(0) = \alpha(T),\ \alpha'(0) = \alpha'(T)$$
$$\alpha''(t) > f(t,\ \alpha(t),\ \alpha'(t))$$
$$\beta''(t) < f(t,\ \beta(t),\ \beta'(t)).$$

b) There exists a positive function $\psi \in C^1[0,\infty)$ such that

$$\int_0^\infty \rho d\rho/\psi(\rho) = +\infty$$

and

$$|f(t,x,x')| < \psi(|x'|)$$

for $|x| \leqslant \max\,[\max \beta(t), -\min \alpha(t)]$.

Then (V.9) - (V.10) has at least one solution.

Remark. Define

$$G_1 = \{(t,x) : \alpha(t) < x < \beta(t)\}.$$

Then G_1 is a CBS relative to (V.9), but not necessarily relative to (V.9.λ).

For $(t_0, x_0) \in \partial G_1$ with $x_0 = \beta(t_0)$ take $V_1(t,x) = x - \beta(t)$.

Then

$$H \begin{bmatrix} 1 \\ y \end{bmatrix} \cdot \begin{bmatrix} 1 \\ y \end{bmatrix} + \text{grad}\, V_1 \cdot \begin{bmatrix} 0 \\ f(t_0,x_0,y) \end{bmatrix} = \begin{bmatrix} -\beta''(t_0) & 0 \\ 0 & 0 \end{bmatrix} \begin{bmatrix} 1 \\ y \end{bmatrix} \cdot \begin{bmatrix} 1 \\ y \end{bmatrix} + \begin{bmatrix} -\beta'(t_0) \\ 1 \end{bmatrix} \begin{bmatrix} 0 \\ f(t_0,x_0,y) \end{bmatrix} =$$

$$= -\beta''(t_0) + f(t_0, x_0, y).$$

If

$$\text{grad } V_1 \cdot \begin{bmatrix} 1 \\ y \end{bmatrix} = -\beta'(t_0) + y = 0,$$

the latter expression becomes

$$-\beta''(t_0) + f(t_0, \beta(t_0), \beta'(t_0)) > 0.$$

However, with λ preceding f there may be a violation of this inequality. Thus we can construct a candidate Ω but there are technical difficulties in applying the Continuation Theorem. The proof below avoids this difficulty by a modification trick which reduces the problem to a special case of Theorem V.7.

Proof of Theorem V.8.

Let R be chosen so that

$$f(t, \beta(t), 0) + R - \beta(t) > 0 \text{ and } R > \beta(t)$$
$$f(t, \alpha(t), 0) - R - \alpha(t) < 0 \text{ and } -R < \alpha(t).$$

Let h(x) be defined as in the proof of Theorem V.7 with N sufficiently large so that $h(x) \geq \max\{\max|\beta'(t)|\max|\alpha'(t)|\}$.

Define

$$F^*(t,x,x') = \begin{cases} f(t,x,h(x)), & x' > h(x), \ |x| \leq R \\ f(t,x,x'), & |x'| \leq h(x), \ |x| \leq R \\ f(t,x,-h(x)), & x' < -h(x), \ |x| \leq R. \end{cases}$$

Define

$$F(t,x,x') = \begin{cases} F^*(t,\beta(t),x') + x - \beta(t), & \beta(t) < x \leq R \\ F^*(t,x,x'), & \alpha(t) \leq x \leq \beta(t) \\ F^*(t,\alpha(t),x') + x - \alpha(t), & -R \leq x < \alpha(t) \end{cases}$$

(Both F and F^* may be extended continuously to all of $[0,T] \times R \times R$).

Note that

$$F(t,R,0) = F^*(t, \beta(t), 0) + R - \beta(t) = f(t, \beta(t), 0) + R - \beta(t) > 0$$

and

$$F(t,-R,0) < 0.$$

For $|x| \leq R$, $|F(t,x,x')|$ is bounded. Thus Theorem V.7 implies that there is at least one solution to

$$x'' = F(t,x,x')$$
$$x(0) = x(T),\ x'(0) = x'(T).$$

with $|x(t)| \leq R$.

Suppose $\max [x(t) - \beta(t)] = x(t_0) - \beta(t_0) > 0$. Then $x'(t_0) = \beta'(t_0)$ and $x''(t_0) - \beta''(t_0) \leq 0$. But

$$x''(t_0) - \beta''(t_0) = F^*(t_0,\beta(t_0),x'(t_0)) + x(t_0) - \beta(t_0) - \beta''(t_0) =$$
$$= f(t_0,\beta(t_0),\beta'(t_0)) + x(t_0) - \beta(t_0) - \beta''(t_0) > 0.$$

Similarly, $\alpha(t) \leq x(t)$. Thus $x(t)$ is a solution to

$$x''(t) = F^*(t,x,x')$$

Suppose $\max [x'(t) - h(x(t))] = x'(t_0) - h(x(t_0)) > 0$. By periodicity we may assume $t_0 \in [0,T)$. Then

$$x''(t_0) - h'(x(t_0))\, x'(t_0) \leq 0.$$

But

$$x''(t_0) - h'(x(t_0))x'(t_0) = f(t_0,x(t_0),h(x(t_0))) + \frac{\psi(h(x(t_0)))}{h(x(t_0))}\, x'(t_0)$$

$$> f(t_0,x(t_0),h(x(t_0))) + \psi(h(x(t_0))).$$

The latter expression is positive by condition b).

Thus $x'(t) \leq h(x(t))$. Sumularly, $x'(t) \geq -h(x(t))$.

<u>Theorem V.9</u>. *Suppose*

a) *Same as Theorem* V.8.

b) *There exist* $\phi(t,x)$, $\psi(t,x) \in C^1([0,T] \times R)$, T-*periodic in* t, *such that* $\phi(t,x) < \psi(t,x)$ and

$$\phi_t(t,x) + \phi_x(t,x)\phi(t,x) \neq f(t,x,\phi(t,x))$$

$$\psi_t(t,x) + \psi_x(t,x)\psi(t,x) \neq f(t,x,\psi(t,x))$$

<u>for</u> $(t,x) \in \{(t,x) : \alpha(t) \leq x \leq \beta(t)\}$.

<u>Then</u> (V.9) - (V.10) <u>has at least one solution</u> $x(t)$ <u>satisfying</u> $\alpha(t) \leq x(t) \leq \beta(t)$ <u>and</u> $\phi(t,x(t)) \leq x'(t) \leq \psi(t,x(t))$.

<u>Proof</u>. Exercise (<u>note</u> : $G = \{(t,x,x') : \alpha(t) < x < \beta(t),\ \phi(t,x) < x' < \psi(t,x)\}$ is a Nagumo set).

<u>Remarks</u>.

1) The strict inequalities in all the hypotheses a) may be removed.

2) The upper and lower solutions required in hypothesis a) of Theorems. V.8 and V.9 are difficult to obtain other than in the case of constants ; i.e., the case of hypothesis a) in Theorem V.7.

4. We now consider

$$x'' = f(t,x,x') \qquad (V.11)$$

$$x(a) = 0 = x(b) \qquad (V.12)$$

where $f : [a,b] \times R \times R \to R$ is continuous.

<u>Theorem V.10</u>. *Under the hypotheses of Theorem* V.7 *the problem* (V.11) - (V.12) *has at least one solution satisfying* $|x(t)| \leq R$.

<u>Proof</u>. Note that $Lx = \lambda\, Nx$ in the formulation V4 is equivalent to (V.9.λ) - (V.12).

We construct G_1 as in the proof of Theorem V.7 ; i.e.,

$$G_1 = \{(t,x) : |x| < R\}$$

is a C B S relative to (V.9.λ) for $\lambda \in (0,1)$. We modify the construction of G as follows. Let $y = h_1(x)$ be the unique solution to

$$\frac{dy}{dx} = -\frac{\psi(y)}{y}$$

$$y(0) = N$$

where N is chosen sufficiently large so that $h_1(x)$ is defined and positive on $[0,R]$. Define $y = h_2(x)$ to be the unique solution to

$$\frac{dy}{dx} = \frac{\psi(y)}{y}$$

$$y(0) = N$$

where N is chosen sufficiently large so that $h_2(x)$ is defined and positive on $[-R,0]$. Define

$$G = \{(t,x,x') : |x| < R, \quad |x'| < h_1(x), \ x \in [0,R], \quad |x'| < h_2(x), \ x \in [-R,0]\}.$$

By arguments analagous to those in the proof of Theorem V.7, G is a Nagumo set relative to (V.9.λ) for $\lambda \in (0,1)$.

Define

$$\Omega = \{x : x \in X, \ (t,x(t),x'(t)) \in G \quad \text{for } t \in [a,b]\}$$

Suppose $x \in \partial\Omega$. By Theorem V.6, $(a,x(a),x(a)) \in \partial G$ or $(b,x(b),x'(b)) \in \partial G$.

Suppose for definiteness that $(a,x(a),x'(a)) \in \partial G$. Then $|x'(a)| = h_1(x(a))$. Suppose for definiteness that $x'(a) = N > 0$. Then

$$u(t) = [x'(t)]^2 - [h_1(x(t))]^2$$

has a relative maximum at $t = a$. Thus $u'(a) \leq 0$. But

$$u'(a) = 2x'(a)x''(a) - 2h_1(x(a))\, h_1'(x(a))x'(a)$$

$$= 2x'(a)[f(a,x(a),x'(a)) + \frac{\psi(x'(a))}{x'(a)} x'(a)] > 0.$$

Other cases are handled in a similar manner thus $x \notin \partial\Omega$.

Clearly the zero function is in Ω, and thus the Continuation Theorem implies existence.

Remark. We used only

$$xf(t,x,y) > -|x|\ \psi(|y|),\ 0 < |x| < R$$

in the proof rather than the full force of condition b).

Theorem V.11. *Suppose*

a) *There exist* $\alpha(t)$, $\beta(t) \in C^2[a,b]$ *such that*

$$\alpha(a) < 0 < \beta(a),\ \alpha(b) < 0 < \beta(b)$$
$$\alpha(t) < \beta(t) \quad ,\ t \in [a,b]$$
$$\alpha''(t) > f(t,\alpha(t),\beta(t)),\ t \in [a,b].$$

b) *Same as Theorem* V.8.

Then (V.11) - (V.12) *has at least one solution satisfying* $\alpha(t) \leq x(t) \leq \beta(t)$.

Proof. By straightforward modification of the proof of Theorem V.8.

5. In this section we seek conditions under which there exist functions $\alpha(t)$ and $\beta(t)$ such that

$$\begin{cases} \alpha''(t) > f(t,\alpha(t),\alpha'(t)), & t \in [a,b] \\ \beta''(t) < f(t,\beta(t),\beta'(t)), & t \in [a,b] \\ \alpha(t) < \beta(t), & t \in [a,b] \\ \alpha(a) < 0 < \beta(a) \\ \alpha(b) < 0 < \beta(b) \end{cases} \qquad (V.13)$$

To construct $\beta(t)$, for example, we suppose

$$f(t,x,x') > -\psi(x,x')$$

and consider the solution $\beta(t)$ to

$$x'' = -\phi(x,x')$$
$$x(a) = M$$
$$x'(a) = 0.$$

Theorem V.12. *Suppose*

$$xf(t,x,x') > -|x|\,\phi(|x'|)$$

for $|x| \geq M_1 > 0$ *where* $\phi \in C^1[0,\infty]$, $\phi(\sigma) > 0$, *and*

$$\int_0^\infty d\sigma / \phi(\sigma) > b - a. \qquad (V.14)$$

Then there exist functions α, β *satisfying* (V.13).

Proof. Let $\beta(t)$ be the unique solution to

$$\beta''(t) = -\phi(|\beta'(t)|)$$
$$\beta(a) = M$$
$$\beta'(a) = 0.$$

Then

$$\beta(t) = M + \int_a^t z(s)ds$$

where $z(t)$ satisfies

$$z' = -\phi(|z(t)|)$$
$$z(a) = 0.$$

By (V.14), $z(t)$ (and hence $\beta(t)$) is defined on $[a,b]$. It is also easily seen that for M sufficiently large $\beta(t) > M_1$ on $[a,b]$. But then

$$\beta''(t) = -\phi(|\beta(t)|) < f(t,\beta(t),\beta'(t)).$$

It is easily verified that we may satisfy (V.13) by taking

$$\alpha(t) = -\beta(t).$$

Remark.

1) We needed to have

$$x\, f(t,x,x') > -|x|\, \phi(|x'|)$$

only on $D = \{(t,x,x') : x \geq M_1 \text{ and } x' \leq 0\} \cup \{(t,x,x') : x \leq -M_1 \text{ and } x' \geq 0\}$.

There are other sets D which can be used if $\beta(t)$ is taken to satisfy different initial conditions.

2) If we take $\beta(t)$ to satisfy

$$\beta''(t) = -\phi(|\beta'(t)|)$$
$$\beta(t_0) = M \qquad (t_0 = a + \tfrac{b-a}{2})$$
$$\beta'(t_0) = 0$$

then we need only

$$\int_0^\infty d\sigma / \phi(\sigma) > \frac{b-a}{2}$$

in place of (V.14).

Corollary V.13. If

a) $f(t,x_1,y) \leq f(t,x_2,y)$ for $x_2 \geq x_1$, and
b) $|f(t,x,y_1) - f(t,x,y_2)| \leq K|y_1 - y_2|$,

then there exist $\alpha(t)$, $\beta(t)$ satisfying (V.13).

Proof. For $x \geqslant 0$,

$$f(t,x,x') \geqslant f(t,0,x') \geqslant f(t,0,0) - K|x'| \geqslant -A - K|x'|$$

where $|f(t,0,0)| \leqslant A$ on $[a,b]$.

For $x \leqslant 0$,

$$f(t,x,x') \leqslant A + K|x'|.$$

we may take $\phi(\rho) = A + K\rho$.

Remark. Under the conditions of Corollary V.13, hypothesis b) of Theorem V.11 is also satisfied. Thus the existence of solutions to the Picard problem is assured.

Example. We show that

$$x'' = e^{x}[1 + (x')^2]$$

$$x(a) = x(b) = 0$$

has a solution.

a) For $x > M_1 > 0$,

$$f(t,x,x') > 0$$

For $x < -M_1 < 0$,

$$f(t,x,x') \leqslant e^{-M_1}[1 + (x')^2].$$

If we take

$$\phi(\rho) = e^{-M_1}(1 + \rho^2),$$

we have

$$\int_0^\infty \frac{d\rho}{\phi(\rho)} = e^{M_1} \int_0^\infty \frac{d\rho}{1 + \rho^2} > b$$

for M_1 sufficiently large.

b) For $|x| \leqslant R$ (arbitrary R) we have

$$|f(t,x,x')| \leqslant e^{R}[1 + (x')^2].$$

If we take

$$\psi(\sigma) = e^R[1 + \sigma^2],$$

then,

$$\int_0^\infty \frac{\sigma d\sigma}{\psi(\sigma)} = \int_0^\infty \frac{\sigma d\sigma}{e^R[1 + \sigma^2]} = + \infty$$

By Theorem V.11 the problem has a solution.

Theorem V.14. *Suppose*

$$x f(t,x,x') > - |x|[A + C |x|^\delta + B|x'|]$$

for $|x| \geq M_1$, *where* $A > 0$, $B,C \geq 0$, *and* $0 \leq \delta < 1$. *Then there exist* $\alpha(t)$, $\beta(t)$ *satisfying* (V.13).

Proof. Define $\beta(t)$ to be the unique solution to

$$\beta''(t) = - A - C|\beta(t)|^\delta - B|\beta'(t)|$$
$$\beta(a) = M$$
$$\beta'(a) = 0.$$

Then

$$\beta(t) = M - \int_0^t e^{B\sigma} \int_0^\sigma [A + C|\beta(s)|^\delta] e^{-Bs} ds\, d\sigma.$$

Then if $M > 1$,

$$\beta(t) \geq M - (A + C M^\delta) \int_0^t e^{B\sigma} \int_0^\sigma e^{-Bs} ds\, d\sigma.$$

and for M sufficiently large $\beta(t) \geq M_1$ on $[a,b]$ and

$$\beta''(t) < f(t,\beta(t),\beta'(t)).$$

Let $\alpha(t) = - \beta(t)$.

Theorem V.15. *Suppose*

$$x f(t,x,x') > - |x|[A + C|x| + B|x'|]$$

for $|x| \geq M_1$ *where* $A, B, C > 0$, $2\Gamma(B,C) > b - a$, and

$$\Gamma(B,C) = \begin{cases} 2D^{-1/2} \tanh^{-1}(\sqrt{D}/B), & D = B^2 - 4C > 0 \\ 2(-D)^{-1/2} \tan^{-1}(\sqrt{-D}/B), & D < 0 \\ 1/2\, B, & D = 0, \end{cases}$$

then there exist $\alpha(t)$, $\beta(t)$ *satisfying* (V.13).

Proof. The result follows from solving

$$\beta'' = -A - B|\beta'(t)| - C|\beta(t)|$$
$$\beta(t_0) = M$$
$$\beta'(t_0) = 0$$

in the three possible cases depending on the roots of $r^2 - Br + C = 0$. The details are left to the reader.

Remarks.

1) Theorems V.12, V.14, and V.15 can be converted to existence theorems for the Picard problem by adding the hypothesis b) of Theorem V.8.

2) If $M_1 = 0$ it is unnecessary to add the additional hypothesis. In this case we need only take $\psi(\sigma) = \phi(|R,\sigma|)$.

3) The existence theorem corresponding to Theorem V.15 is sharp in the sense that if we apply the theorem to

$$x'' = -x + f(t)$$
$$x(a) = x(b) = 0,$$

then

$$x[-x + f(t)] > -|x|[A + |x|]$$

where

$$\max_{[a,b]} |f(t)| \leq A.$$

Thus the condition

$$\Gamma(B,C) > \frac{b-a}{2}$$

becomes

$$b - a < \pi.$$

We know that for some choices of f(t), if $b - a \geq \pi$, then the problem has no solution.

Theorem V.16. *Suppose*

$$x\, f(t,x,x') > -|x|\, \phi(|x|, |x'|)$$

for $|x| \geq M_1$ *where* $\phi(\rho,\sigma) \in C^1(R^2)$. *Let* $\beta(t,M)$ *be the unique solution to*

$$x'' = -\phi(|x|, |x'|)$$
$$x(a) = M$$
$$x'(a) = 0.$$

Suppose $\beta(t,M) \to +\infty$ *uniformly in* t *on* [a,b] *and* $\beta(t,M) \geq M_1$ *for* $M \geq M_0$. *Then any solution to the Picard problem satisfies*

$$|x(t)| < \beta(t, M_0), \qquad t \in [a,b].$$

Proof. Suppose not. Suppose there exists x(t) such that

$$x(t_0) \geq \beta(t_0, M_0).$$

Since $\beta(t,M) \to +\infty$ uniformly on [a,b], there exists $\tilde{M} \geq M_0$ and $\tilde{t} \in (a,b)$ such that

$$\beta(\tilde{t},\tilde{M}) = x(\tilde{t})$$
$$\beta(t,M) \geq x(t) \text{ on } [a,b].$$

But since $\beta(t,\tilde{M}) \geq M_1$, $\tilde{t} \neq a,b$ and we must have

$$\beta'(\tilde{t},\tilde{M}) = x'(\tilde{t})$$
$$\beta''(\tilde{t},\tilde{M}) = x''(\tilde{t}).$$

But we also have the contradictory inequality

$$\beta''(\tilde{t},\tilde{M}) = -\phi(|\beta(\tilde{t},\tilde{M})|), |\beta'(\tilde{t},\tilde{M})|) < f(\tilde{t}, \beta(\tilde{t}), \beta'(\tilde{t})) = x''(\tilde{t},\tilde{M}).$$

By similar arguments, $x(t) > -\beta(t,M_0)$.

Remarks.

1) This theorem is the first which yields a bound on all possible solutions to the problem.

2) The hypotheses of Theorems V.12, V.14, and V.15 yield $\beta(t,M)$ as required in Theorem V.16.

3) The hypotheses could be formulated in terms of a family of upper and lower solutions depending continuously on a parameter.

4) Under the hypotheses of Theorem V.16 we also have a priori bounds for other boundary value problems.

If $x(t)$ is a solution to (V.9) on $[a,b]$:

$$|x(a)|, |x(b)| < M_1 \implies |x(t)| < \beta(t,M_0) \leq M_0$$

$$x'(a) = 0, |x(b)| < M_1 \implies |x(t)| < \beta(t,M_0) \leq M_0$$

$$|x(a)| \leq M_1, x(b) = 0 \implies |x(t)| \leq \beta(b - (t-a), M_0) \leq M_0$$

The proofs are nearly identical to the proof of Theorem V.16.

Theorem V.17. *Suppose*

a) *There exists* ϕ *so that the hypotheses of Theorem* V.16 *are satisfied (with respect to* $[0,T]$*).*

b) $f(t,R,0) < 0$, $f(t,-R,0) > 0$ *for* $R \geq M_1$.

c) *There exists* $\psi > 0$, $\psi \in C^1[0,+\infty)$ *such that*

$$\int_0^\infty \rho d\rho \,/\, \psi(\rho) = +\infty$$

and

$$|f(t,x,x')| < \psi(|x'|)$$

for $|x| \leq M_0$ *(where* M_0 *is given by Theorem* V.16*).*

Then the periodic problem has at least one solution.

<u>Proof</u>. Let $\beta(t,M_0)$ be given by Theorem V.16. Let $h(x)$ be constructed as in the proof of Theorem V.7 with $R = M_0$ and let

$$G = \{(t,x,x') : |x| < M_0,\ |y| < h(x)\}$$

$$\Omega = \{x \in X : (t,x(t),x'(t)) \in G,\ t \in [0,T]\}.$$

a) Suppose $x \in \partial\Omega$ and x is a solution to (V.9.λ) - (V.10). By the argument in the proof of Theorem V.7, $|x(t)| < h(x(t))$ on $[0,T]$. We must have $|x(t_0)| = M_0$ for some t_0. We suppose for definiteness that

$$\max_{[0,T]} x(t) = x(t_0) = M_0.$$

Now $x(t)$ is nonconstant from b) and also from b) we deduce that the minimum of $x(t)$ must occur at some point t_1 where $0 < x(t_1) < M_1$. We assume for definiteness that $t_1 > t_0$. By periodicity of $x(t)$, $x(t)$ satisfies the boundary conditions

$$x'(t_0) = 0,\ |x(t_1)| < M_1.$$

It is easily seen from Remark 4 above that

$$x(t) < \beta(t - t_0, M_0) \leq M_0,\ t \in [t_0, t_1].$$

But then we contradict the definition of t_0.

b) The verification that hypothesis b) of the Continuation Theorem is satisfied is identical to the corresponding part of the proof of Theorem V.7 except that R and -R are reversed.

6. We return now to consideration of the system (V.1)

$$x'' = f(t,x,x').$$

We will have need to refer to

$$x'' = \lambda f(t,x,x'). \qquad (V.15.\lambda)$$

In Sections 3 and 4 we exploited curvature bound sets where the curvature of the boundary varied over time (or not at all). We now consider CBS 's where the curvature of the boundary varies only with respect to the space variables. For this purpose, we introduce an autonomous version of the curvature bound set.

Definition V.18. A set $\hat{G}_1$ will be called an autonomous CBS if $G_1 = [a,b] \times \hat{G}_1$ is a CBS and for each $(t_0,x_0) \in \partial G_1, V_1(t_0,x_0;t,x) = V_1(x_0;x)$ can be taken independent of t_0 and t.

Note. In this case (V.8) may be written more simply as

$$V_{1xx}y \cdot y + V_{1x} \cdot f(t,x_0,y) > 0, \text{ for } y \text{ such that } V_{ix} \cdot y = 0 \qquad (V.16)$$

where V_{1xx} denotes the Hessian and V_{1x} denotes the gradient of V_1 at x_0.

Theorem V.19. *Suppose*

a) $\hat{G}_1$ *is an autonomous* CBS *relative to* (V.1) *on* [0,T].

b) $\hat{G}_1$ *is convex.*

c) $0 \in \hat{G}_1$.

d) V_{1xx} *is positive semidefinite for* $x_0 \in \partial \hat{G}_1$.

e) *There exists a positive function* $\psi \in C^1[0,\infty)$ *such that*

$$|f_i(t,x,y)| \leq \psi(|y_i|) \quad \text{for} \quad x \in \hat{G}_1, \quad i = 1,2,\dots,n \;;$$

and

$$\int_0^\infty \frac{\sigma d\sigma}{\psi(\sigma)} = +\infty .$$

Then the periodic problem for (I.1) *has at least one solution.*

Proof.

a) Since V_{1xx} is positive semidefinite

$$V_{1xx}\, y \cdot y + V_{1x} \cdot \lambda\, f(t,x_0,y) > 0$$

for y such that grad $V_1 \cdot y = 0$, and $\hat{G}_1$ is a CBS relative to (V.15.λ) on [0,T] for $\lambda \in (0,1)$. Construct $h(\rho)$ as in the proof of Theorem V.1 with

$$R = \max\{|x_i| : x \in \hat{G}_1\},$$

that is, $h(\rho)$ is the unique solution to

$$\frac{d\sigma}{d\rho} = -\frac{\psi(\sigma)}{\sigma}$$

$$\sigma(-R) = N$$

where N is chosen sufficiently large that $\sigma = h(\rho) > 0$ on $[-R,R]$. Define

$$G = \{(t,x,y) : x \in \hat{G}_1 ,\ |y_i| < h(x_i)\}$$

$$\Omega = \{x \in X : (t,\ x(t),\ x'(t)) \in G \text{ for } t \in [0,T]\}.$$

Suppose $(t_0, x_0, y_0) \in \partial G$ with $t_0 \in (a,b)$ and $(t_0, x_0) \notin \partial G_1$. Then $|y_{io}| = h(x_{io})$ for some i. Define

$$V(t,x,y) = y_i^2 - (h(x_i))^2 .$$

Then i), ii), iii) of Definition V.1 are satisfied. Moreover,

$$\text{grad } V \cdot \begin{vmatrix} y_0 \\ \lambda f(t_0,x_0,y_0) \end{vmatrix} = -2h(x_{io})h'(x_{io})y_{io} + 2y_{io}\lambda f_i(t_0,x_0,y_0)$$

$$= 2y_{io}\Big[|y_{io}|\frac{\psi(|y_{io}|)}{|y_{io}|} + \lambda f_i(t_0,x_0,y_0)\Big].$$

This latter quantity is nonzero by hypothesis e). Thus iv) of Definition V.1 is satisfied and Definition V.5 is satisfied ; i.e., G is a Nagumo set relative to (V.15.λ) for $\lambda \in (0,1)$.

Suppose $x \in \partial\Omega$ and x is a solution to the periodic problem for (V 15.λ). By Theorem V.6 this can happen only if $(0,\ x(0),\ x'(0))$ and

$(T, x(T), x'(T)) \in \partial G$. Then by periodicity, one of the following must hold :

a) $|x_i'(0)| = h(x_i(0))$; $|x_i'(T)| = h(x_i(T))$, for some i

b) $x(0) = x(T) \in \partial \hat{G}_1$.

In the first case we may argue exactly as in the proof of Theorem V.7 to reach a contradiction ; i.e., consider

$$\gamma(t) = (x_i'(t))^2 - (h(x_i(t)))^2.$$

We have $\gamma(t) \leqslant 0$ on $[0,T]$, $\gamma(0) = \gamma(T) = 0$. Thus $\gamma'(0) \leqslant 0$, $\gamma'(T) \geqslant 0$. But

$$\gamma'(t) = 2x_i'(t)\, x_i''(t) - 2h(x_i(t))\, h'(x_i(t))\, x_i'(t)$$
$$\gamma'(0) = 2x_i'(0)\, [\lambda f_i(0, x(0), x'(0)) + \psi(|x'(0)|)]$$
$$\gamma'(T) = 2x_i'(T)\, [\lambda f_i(T, x(T), x'(T)) + \psi(|x'(T)|)].$$

The two expressions on the right are nonzero and have the same sign. In the second case, let $V_1(x)$ be the function associated with $x(0) = x(T)$. Consider

$$u(t) = V_1(x(t)).$$

Then $u(t) \leqslant 0$ on $[0,T]$ and $u(0) = u(T) = 0$. Moreover, by periodicity

$$u'(0) = V_{1x} \cdot x'(0) = V_{1x} \cdot x'(T) = u'(T) = 0$$

Thus we must have $u''(0) \leqslant 0$. But

$$u''(0) = V_{xx}\, x'(0) \cdot x'(0) + V_x \cdot \lambda f(0, x(0), x'(0)) > 0.$$

b) We have

$$\mathrm{Ker}\, L \cap \Omega = \{x : x = c,\ c \in \hat{G}_1\}$$
$$\mathrm{Ker}\, L \cap \partial\Omega = \{x : x = c,\ c \in \partial\hat{G}_1\}$$
$$QN(c) = \frac{1}{T}\int_0^T f(t,c,0)\,dt.$$

We may take $J = I$ (or more precisely, J is the natural mapping from the constant functions as a subspace of into the constant functions as a subspace of x).

From (V.16) if we take $y = 0$, $x = c \in \partial \hat{G}_1$ we have

$$V_{1x} \cdot f(t,c,0) > 0,\ c \in \partial \hat{G}_1.$$

Thus

$$\int_0^T V_{1x} \cdot f(t,c,0)\,dt > 0,\quad c \in \partial \hat{G}_1$$

and

$$\frac{1}{T} V_{1x} \int_0^T f(t,c,0)dt > 0 = V_{1x} \cdot QN(c) > 0, \; c \in \partial \hat{G}_1.$$

Moreover, we claim

$$V_{1x} \cdot c \geq 0, \quad c \in \partial \hat{G}_1.$$

In fact, since $0 \in \hat{G}_1$ and $\hat{G}_1$ is convex, $V_1(\lambda c \; ; \; c) \leq 0$ for $\lambda \in [0,1]$. Since $V_1(1 \cdot c \; ; \; c) = 0$ we must have

$$\left.\frac{d\, V_1(\lambda c;c)}{d\lambda}\right|_{\lambda = 1} = V_{1x} \cdot c \geq 0.$$

Thus

$$\lambda V_{1x} \cdot QN(c) + (1-\lambda)V_{1x} \cdot c > 0, \; c \in \partial \hat{G}_1,$$

$$V_{1x} \cdot [\lambda\, QN(c) + (1-\lambda)c] > 0, \; c \in \partial \hat{G}_1,$$

and

$$[\lambda\, QN(c) + (1-\lambda)c] \neq 0, \; c \in \partial \hat{G}_1.$$

By the Poincare - Bohl Theorem,

$$d[QN(c), \text{Ker } L \cap \Omega, 0] = d[I, \text{Ker } L \cap \Omega, 0] \neq 0.$$

Thus the Continuation Theorem yields the desired result.

Remark. The condition e) is very restrictive. The following lemma presents a simple alternative.

Lemma V.20. *Suppose*

a) $x(t) \in \overline{\hat{G}_1}$ *on* $[a;b]$ *where* $\hat{G}_1$ *is bounded.*

b) $\|x''(t)\| \leq \psi(\|x'(t)\|)$ *on* $[a,b]$ *where* ψ *is positive, continuous, and non-decreasing on* $[0,+\infty)$, *and satisfies*

$$\lim_{s \to +\infty} \frac{s^2}{\psi(s)} = +\infty.$$

Then there exists M *depending only on* $\hat{G}_1$ *and* ψ *such that* $\|x'(t)\| \leq M$ *on* $[a,b]$.

Proof. Suppose

$$\sup_{x\in\hat{G}_1} \|x\| \leq R.$$

Choose M so that

$$M \geq 8\, R/(b - a) \tag{V.17}$$

and

$$s^2/\psi(s) > 4\, R,\ s > M. \tag{V.18}$$

Let

$$q = \max_{t\in[a,b]} \|x'(t)\| = \|x'(t_0)\| > M.$$

Then if $t_0 + h \in [a,b]$ we have

$$x(t_0+h) - x(t_0) = x(t_0) + x'(t_0)h + \int_0^1 x''(t_0+sh)h^2\,(1-s)ds$$

$$q|h| \leq 2R + \int_0^1 \psi(\|x'(t_0+sh)\|)\, h^2(1-s)ds$$

$$q|h| \leq 2R + \psi(q)\, h^2/2$$

$$q \leq \frac{2R}{|h|} + \psi(q)|h|/2.$$

Note that this inequality holds for $|h| \leq (b-a)/2$. Using (V.18)

$$q < \frac{2R}{|h|} + \frac{q^2|h|}{8R}.$$

The expression on the right has a minimum at $|h| = \frac{4R}{q}$. If $q > M \geq 8R/(b-a)$,

$$\frac{4R}{q} \leq \frac{4R(b-a)}{8R} = (b-a)/2,$$

and

$$q < q/2 + q/2 = q. \quad \text{Thus} \quad q \leq M.$$

Theorem V.21. *Suppose*

a) $\hat{G}_1$ *is as in hypotheses* a), b), c) *and* d) of *Theorem* V.19.

b) *There exists* ψ *as in Lemma* V.20 *with*

$$\|f(t,x,y)\| \leq \psi(\|y\|),\quad x \in \overline{\hat{G}_1}.$$

Then the periodic problem for (V.1) *has at least one solution.*

Proof. Let

$$\Omega = \{x \in X : x(t) \in \hat{G}_1,\ \|x'(t)\| < M + 1 \text{ on } [0,T]\}$$

where M is given by Lemma V.20.

Suppose $x \in \partial\Omega$ and x is a solution to the periodic problem for (V.15.λ). By Lemma V.20, $\|x'(t)\| \leq M$, and since $\hat{G}_1$ is a CBS, Theorem V.4 implies that $x(0) = x(T) \in \partial \hat{G}_1$. This yields a contradiction just as in the proof of Theorem V.19.

The remainder of the proof is identical to the proof of Theorem V.19.

Note : If $\hat{G}_1$ is convex, then for any $x_0 \in \partial G_1$ there is an outer normal $n(x_0)$ which has the property that

$$\hat{G}_1 \subset \{x : n(x_0) \cdot (x-x_0) < 0\}.$$

Corollary V.22. *Suppose*

a) $\hat{G}_1$ *is convex with* $0 \in \hat{G}_1$.

b) *For* $x_0 \in \partial \hat{G}_1$ *there exists an outer normal* $n(x_0)$ *to* $\partial \hat{G}_1$ *at* x_0 *such that if* $n(x_0) \cdot y = 0$,

$$n(x_0) \cdot f(t,x_0,y) > 0$$

c) *There exists* ψ *as in Lemma* V.20 *with*

$$\|f(t,x,y)\| \leq \psi(\|y\|), \quad x \in \bar{\hat{G}}_1.$$

Then the periodic problem for (V.1) *has at least one solution.*

Proof. For $x_0 \in \partial G_1$ take

$$V_1(x;x_0) = n(x_0)\,(x-x_0).$$

Then $V_1(x;x_0)$ clearly satisfies hypotheses i), ii) and iii) of Definition V.3. Moreover,

$$V_{1xx}\, y \cdot y + V_{1x} \cdot f(t,x_0,y) = 0 + n(x_0)\cdot f(t,x_0,y) > 0$$

whenever $V_{1x} \cdot y = n(x_0)y = 0$ by hypothesis b). Thus $\hat{G}$ is an autonomous CBS.

Since $V_{1xx} = 0$ is positive semidefinite it is seen that all the hypotheses of Theorem V.21 are satisfied.

Corollary V.23. *Suppose*

a) $\|y\|^2 + x \cdot f(t,x,y) > 0$ *when* $\|x\| = R$ *and* $x \cdot y = 0$

b) *There exists* ψ *as in Lemma* V.20 *with*

$$\|f(t,x,y)\| \leq \psi(\|y\|), \quad x \in \overline{\hat{G}_1}$$

Then the periodic problem for (V.1) *has at least one solution.*

Proof. Take $\hat{G}_1 = B_R(0)$; i.e., a ball of radius R centered at 0. For $x_0 \in \partial\hat{G}_1$ take

$$V_1(x) = \frac{1}{2}\|x\|^2 - \frac{1}{2}R^2.$$

We have

$$V_{1x} = x$$

$$V_{1xx} = I$$

$$V_{1xx} \quad y \cdot y + V_{1x} \cdot f(t,x_0,y) = \|y\|^2 + x \cdot f(t,x_0,y)$$

By hypothesis a) the latter expression is positive when $x_0 \in \partial G_1$; i.e., $\|x_0\| = R$, and $V_{1x} \cdot y = x_0 \cdot y = 0$. Thus $\hat{G}_1$ is an autonomous CBS. Since $V_{1xx} = I$ is positive semidefinite, all of the hypotheses of Theorem V.21 are satisfied.

Theorem V.24. *Under the hypotheses of Theorems* I.19 *and* I.21 *and Corollaries* I.22 *and* I.23, *the Picard problem for* (V.1) *also has at least one solution.*

Proof. Left to the reader.

Remark. The reader can no doubt recognize possibilities for obtaining existence theorems employing curvature bound sets with the curvature at boundary points varying with both t and x. We limit ourselves to one Theorem and Corollary of this type.

Theorem V.25. *Suppose*

a) *There exists a positive, real valued function* $\beta(t) \in C^2[a,b]$ *and a real valued function* $W(x) \in C^2(R^n)$ *such that*

$$\beta''(t) < 0 \quad on\ [a,b]$$

$$W_{xx}\, y \cdot y + W_x \cdot f(t,x,y) > \beta''(t)$$

for (t,x,y) *such that*

$$W(x) = \beta(t)$$
$$W_x \cdot y = \beta'(t).$$

b) $G_1 = \{(t,x) : W(x) - \beta(t) < 0\}$ *is bounded and* $(t,0) \in G_1$ *for* $t \in [a,b]$.

c) W_{xx} *is positive semi-definite for* $x \in R^n$.

d) *There exists* ψ *as in Lemma* V.20 *with*

$$\|f(t,x,y)\| \leq \psi(\|y\|), \quad (t,x) \in \overline{G}_1 .$$

Then the Picard problem has at least one solution.

Proof. Define

$$G = \{(t,x,y) : (t,x) \in G_1, \quad \|y\| < M + 1\}$$
$$\Omega = \{x \in X : (t, x(t), x'(t)) \in G\}$$

where M is given by Lemma V.20.

For $(t_0, x_0) \in \partial G_1$, let

$$V_1(t,x) = W(x) - \beta(t).$$

We have

$$H \begin{bmatrix} 1 \\ y \end{bmatrix} \cdot \begin{bmatrix} 1 \\ y \end{bmatrix} + \text{grad } V_1 \cdot \begin{bmatrix} 0 \\ \lambda f(t_0,x_0,y) \end{bmatrix}$$

$$= W_{xx}\, y \cdot y - \beta''(t_0) + \lambda W_x \cdot f(t,x_0,y_0).$$

Moreover,

$$\text{grad } V_1 \cdot \begin{bmatrix} 1 \\ y \end{bmatrix} = W_x\, y - \beta'(t\ ).$$

Since W_{xx} is positive semidefinite, $\beta''(t_0) < 0$, and $W(x_0) = \beta(t_0)((t_0,x_0) \in \partial G_1)$ hypothesis a) implies that G_1 is a CBS relative to (V.15.λ) for $\lambda \in (0,1)$.

Let $x(t)$ be a solution to the Picard problem for (V.15.λ). Since $(a, x(a)), (b, x(b)) \in G_1$, Theorem V.4 and Lemma V.20 imply that $x \notin \partial\Omega$. The Continuation Theorem then applies.

Corollary V.26. *Suppose*

a) *There exists* $K > 0$ *such that*

$$\|y\|^2 + x \cdot f(t,x,y) \geq - K[1 + \|x\| + (x \cdot y)]$$

b) *For any bounded set* $\hat{G}_1$ *there exists a* ψ *as in Lemma* V.20 *with*

$$\|f(t,x,y)\| \leq \psi(\|y\|), \qquad x \in \hat{G}_1.$$

Then the Picard problem has at least one solution.

Proof. Take $W(x)$ and $\beta(t)$ so that

$$W(x) = \|x\|^2/2$$

$$\beta''(t) = - K[1 + (2|\beta(t)|)^{1/2} + |\beta'(t)|].$$

(By Theorem V.14 we know that there exists a $\beta(t)$ satisfying the equation with $\beta(t) > 0$ on $[a,b]$). Note that

$$\begin{aligned} W_{xx}\, y \cdot y + W_x \cdot f(t,x,y) &= \|y\|^2 + x \cdot f(t,x,y) \\ &\geq - K[1 + \|x\| + |x \cdot y|] \\ &\geq \beta''(t) \end{aligned}$$

if $\qquad W(x) = \|x\|^2/2 = \beta(t)$ and $W_x \cdot y = x \cdot y = \beta'(t)$.

The remaining hypotheses of Theorem V.25 are easily verified.

7. The continuation method may be used to obtain existence theorems for various classes of problems involving more general boundary conditions and more general equations.

We consider the problem

$$x' = f(t,x) \qquad (V.19)$$

$$0 = g(x(a),x(b)) \qquad (V.20)$$

where

$$f : [a,b] \times R^n \to R^n$$

and

$$g : R^n \times R^n \to R^n$$

are continuous.

We formulate this problem in a function space setting as follows. Let

$$X = C[a,b]$$
$$Z = C[a,b] \times R^n$$
$$\text{dom } L = C^1[a,b]$$

$$L : \text{dom } L \to Z, \qquad x \longmapsto (x',0)$$
$$N : X \to Z, \qquad x \longmapsto (f(\cdot, x(\cdot)), g(x(a), x(b)).$$

(Note that these more general boundary conditions are not incorporated into the domain of L). We have

$$\text{Ker } L = \{x : x = c \in R^n \quad \text{for} \quad t \in [a,b]\}$$
$$\text{Im } L = C[a,b] \times \{0\}$$
$$\dim \text{Ker } L = n = \text{codim Im } L.$$

Since Im L is clearly closed, L is Fredholm of index 0. We may take

$$P\,x = x(a),$$
$$Qz = Q(y,d) = (0,d),$$

and

$$K_P\, z = \int_a^t y(s)ds.$$

Then

$$K_P(I-Q)Nx = \int_a^t f(s,x(s))ds.$$

It is easily seen that N is L-compact on $\overline{\Omega}$ for any open bounded set Ω.

The definition of bound set in Section 2 is immediately generalizable to (V. 19). For simplicity we omit the dependence of the set on t.

Definition V.27. An open bounded subset G of R^n will be called a bound set relative to (V.19) on [a,b] if for any $x_0 \in \partial G$ there exists $V(x;x_0) = V(x)$ such that: i) $V \in C^1(R^n)$

ii) $V(x_0) = 0$

iii) $G \subset \{X : V(x) < 0\}$

iv) $\operatorname{grad} V(x_0) \cdot f(t,x_0) \neq 0, \quad t \in [a,b]$.

Theorem V.28. Suppose

a) G is a bound set relative to (V. 19)

b) If x is a solution to

$$x' = \lambda f(t,x), \qquad (\lambda \in (0,1)) \tag{V.21}$$

$$0 = g(x(a), x(b)), \tag{V.22}$$

then $(x(a), x(b)) \notin \partial G$.

c) $g(c,c) \neq 0, \qquad c \in \partial G$

d) $d[g(c,c),G,0] \neq 0$.

Then (V.19) - (V.20) has at least one solution x(t) such that $x(t) \in \overline{G}$ for $t \in [a,b]$.

Proof. Let

$$\Omega = \{x \in X : \quad x(t) \in G \text{ for } t \in [a,b]\}.$$

Let x(t) be a solution to (V.21) - (V.22) for some $\lambda \in (0,1)$.

If $x \in \partial\Omega$ then, $x(t_0) \in \partial G$ for some $t_0 \in [a,b]$.

By hypothesis b), $t_0 \in (a,b)$. Let $V(x) = V(x;x_0)$ be the function given by Definition V.27. Then

$$V(x(t)) \leqslant 0, \ t \in [a,b]$$

$$V(x(t_0)) = 0.$$

Thus

$$\frac{d[V(x(t))]}{dt}\Big|_{t=t_0} = \text{grad } V(x(t_0)).\lambda f(t_0,x(t_0)) = 0.$$

This contradicts iv) in Definition V.27.

Corollary V.29. Suppose there exist $R_i > 0$, $i= 1,2, \ldots, n$ such that

$$f_i\ (t,X(\pm R_i)) \neq 0 \tag{V.23}$$

$$g_i\ (X(R_i))g_i(X(-R_i)) < 0 \tag{V.24}$$

for $X(\pm R_i) = (x_1, \ldots, x_{i-1}, \pm R_i, x_{i-1}, \ldots, x_n)$, $|x_j| \leqslant R_j$.

Then the problem

$$x' = f(t,x)$$

$$g(x(a),x(b)) = \begin{bmatrix} g_1(x(a)) \\ g_2(x(a)) \\ \vdots \\ g_K(x(a)) \\ g_{K+1}(x(b)) \\ \vdots \\ g_n(x(b)) \end{bmatrix} = 0 \tag{V.25}$$

has at least solution $x(t)$ satisfying $|x_i(t)| \leqslant R_i$, $i= 1,2, \ldots,n$.

Proof. We define

$$G = \{(t,x) : |x_i| < R_i\}.$$

If $x_0 \in \partial G$. Then

$$x_0 = X_0(\pm R_i) = (x_1^0, \ldots, x_{i-1}^0, \pm R_i, x_{i+1}^0, \ldots, x_n^0), \ |x_j^0| \leqslant R_j$$

for some i. Choose for definiteness, the plus sign.

We define

$$V(x;x_0) \equiv x_i - R_i.$$

From (V.23) the requirements of Definition V.27 are easily seen to be satisfied and G is a bound set.

Suppose x(t) is a solution to (V.19)-(V.25). If x(a) or x(b) $\in \partial G$ (I.24) is contradicted. The condition (V.24) implies that $g(c,c) \neq 0$ for $c \in \partial G$ and by Miranda's theorem

$$d[g(c,c),G,0] = \pm 1.$$

Thus by Theorem V.28, the corollary is proved.

Remark. Other corollaries to Theorem V.28 can be given and examples may be constructed where $d[g(c,c),G,0] \neq 1$.

We consider

$$x' = f(t,x) \qquad (V.26)$$

$$x(0) = x(T) \qquad (V.27)$$

Rather than employ the operator structure introduced earlier in this section we formulate (V.26)-(V.27) as an operator equation similar to that developed for second order periodic problems. Let

$$X = \{x : x(0) = x(T)\} \cap C[0,T]$$

$$Z = C[0,T]$$

$$\text{dom } L = C'[0,T] \cap X$$

$$L : X \to Z,\ x \to x'$$

$$N : X \to Z,\ x \to f(\cdot,x(\cdot)).$$

We then have

$$\text{Ker } L = \{x : x = C \quad \in R^n\}$$

and

$$\text{Im } L = \{z : \int_0^T z(s)\, ds = 0\}.$$

The latter follows since $x' = z$ if and only if

$$x(t) = \int_0^t z(s)\, ds + d,$$

and $x(0) = x(T)$ implies that

$$d = \int_0^T z(s)\, ds + d.$$

It is easily seen that

$$\dim \text{Ker } L = \text{codim Im } L = n.$$

Since Im L is closed, L is Fredholm of index 0.

We take

$$Px = x(a)$$

$$Qz = \frac{1}{T} \int_0^T z(s)\, ds.$$

$$K_p z = \int_0^t z(s)\,ds .$$

Again it is easily seen by means of the Ascoli-Arzela Theorem that

$$\overline{K_p\ (I-Q)\ N(\Omega)}$$

is compact for any open bounded Ω. Since N is continuous, N is L-compact

Theorem V.30. Suppose

a) G is a bound set relative to (V.26) (on $[0,T]$) with $0 \in G$.

b) $QN(c) \neq 0$ for $c \in \partial G$ and $d[QN,G,0] \neq 0$.

Then (V.26)-(V.27) has at least one solution $x(t)$ such that $x(t) \in \overline{G}$ for $t \in [0,T]$.

Proof. Let

$$\Omega = \{ x \in X : x(t) \in G \text{ for } t \in [0,T]\}.$$

Suppose X is a solution to $Lx = \lambda Nx$, $\lambda \in (0,1)$.

If $x \in \partial\Omega$, then $x(t_0) \in \partial G$ for some $t_0 \in [0,T]$.

Let $V(x;x(t_0)) = V(x)$ be the function given by Definition V.27.

Then

$$V(x(t)) \leq 0,\ t \in [0,T]$$

and

$$V(x(t_0)) = 0.$$

We have

$$\left.\frac{d[V(x(t))]}{dt}\right|_{t=t_0} = \operatorname{grad} V(x(t_0)).\lambda f(t_0,x(t_0)).$$

If $t_0 \in (0,T)$, then we must have $\left.\frac{d[V(x(t))]}{dt}\right|_{t=t_0} = 0$

which contradicts Definition V.27. If $t_0 = 0$ (or $t_0 = T$), then

$$\left.\frac{d[V(x(t))]}{dt}\right|_{t=0} = \text{grad } V(x(0)).\ \lambda f(0,x(0)) \leq 0$$

and

$$\left.\frac{d[V(X(t))]}{dt}\right|_{t=T} = \text{grad } V(x(T)).\lambda f(T,x(T)) \geq 0.$$

But then since $X(0) = X(T)$, by continuity there must exist $t_1 \in [a,b]$ such that

$$\text{grad } V(x(0)).\lambda\ f(t_1,\ x(0)) = 0$$

again contradicting Definition V.27. Thus $x \in \partial\Omega$ is contradicted.
Since we may take J: Im $Q \to$ Ker L to be essentially an identity isomorphism, the conclusion of Theorem V.30 follows the Continuation Theorem.

Corollary V.31. *Suppose* G *is an open, bounded convex subset* G *of* R^n *such that* :

a) $0 \in G$

b) *For each* $x_0 \in \partial G$ *there exists an outer normal* $n(x_0)$ *with*

$$n(x_0).f(t,x_0) > 0 \text{ (\textit{respectively} } < 0).$$

Then (V.26)-(V.27) *has at least one solution* $x(t)$ *such that* $x(t) \in G$ *for* $t \in [0,T]$.

Proof. Note that for any such convex set G and for any $x_0 \in \partial G$ there exists at least one outer normal $n(x_0)$; i.e., a vector $n(x_0) \neq 0$ such that

$$n(x_0).x_0 > 0$$

and

$$G \subset \{x : (x-x_0) . n(x_0) < 0\}.$$

For any such $x_0 \in \partial G$ let $V(x;x_0) = (x-x_0).n(x_0)$.

By hypothesis b) the conditions of Definition V.27 are satisfied and G is a bound set.

For $t \in [0,T]$ and $x \in \partial G$

$$n(x).f(t,x) > 0 \text{ (resp. } < 0).$$

Thus

$$n(c).QN(c) > 0 \text{ (resp } < 0), \; c \in \partial G$$

and

$$n(c).[(1-\sigma)c + \sigma \, QN(c)] > 0, \; c \in \partial G$$

$$\{n(c).[-(1-\sigma)c + \sigma QN(c)] < 0\}.$$

But then

$$(1-\sigma)c + \sigma QN(c) \neq 0$$

$$\{-(1-\sigma)c + \sigma \, QN(c) \neq 0\}.$$

Thus

$$d[QN(c), G, 0] = d[c, G, 0] = 1 \; \{d[-c, G, 0] = (-1)^n\}.$$

Theorem V.30 then implies that (V.26)-(V.27) has at least one periodic solution

An existence theorem may also be formulated in terms of "guiding functions".

Definition V.32. A function $W : R^n \to R$ is a guiding function for equation (V.26) on $[0,T]$ if $V \in C^1(R^n)$ and there exists $R > 0$ such that for $t \in [0,T]$ and $\|x\| \geq R$

$$\text{grad } W(x).f(t,x)$$

is of constant sign.

Corollary V.33. *If $W(x)$ is a guiding function for equation (V.26) on $[0,T]$ such that*

$$|W(x)| \to +\infty \text{ as } \|x\| \to +\infty, \tag{V.28}$$

then problem (V.26) - (V.27) has at least one solution.

Proof. Suppose for definiteness that

$$W(x) \to +\infty \text{ as } \|x\| \to +\infty$$

and

$$\text{grad } W(x).f(t,x) > 0, \; t \in [0,T], \; \|x\| \geq R.$$

Let

$$\rho > \max \{ W(x) : \|x\| \leqslant R\},$$

and

$$G \equiv \{x : W(x) < \rho\}.$$

Then G is certainly open and by (V.28) G is bounded.

Suppose $x_0 \in \partial G$. Then by the definition of G, $\|x\| > R$.

If we take

$$V(x; x_0) \equiv W(x) - \rho,$$

then the definition of bound set is satisfied.

Since

$$\text{grad } W(x).f(t,x) > 0, \quad t \in [0,T], \ \|x\| \geqslant R,$$

$$\text{grad } W(c).\frac{1}{T}\int_0^T f(t,c)\, dt > 0, \qquad c \in \partial G.$$

or

$$\text{grad } W(c).QN(c) > 0, \qquad c \in \partial G.$$

It follows from a corollary to the theorem of Poincare-Bohl (See Rouche and Mawhin, Equations Différentielles Ordinaires, tome 2, p. 179) that

$$d[QN(c), G,0] = d\,[\text{grad } W(c),G,0]$$

Moreover, by the theorem of Kranosel'skii and the property of excision, grad $W(c) \neq 0$ for $\|c\| \geqslant R$ implies

$$d[\text{grad } W(c),G,0] = d[\text{grad } W(c), B_R(0),0] = 1.$$

Remark. We note in passing that our techniques apply to classes of problems which are formulated in other ways. For example the classes :

$$x^{(m)} = f(t,x, \ldots, x^{(m-1)})$$

$$g(x(a),x(b),x'(a),x'(b), \ldots, x^{(m-1)}(a), x^{(m-1)}(b)) = 0$$

$$(f : [a,b] \times R^{nm} \to R^n \text{ and } g : R^{2mn} \to R^{mn})$$

and $$x' = f(t,x)$$
$$Ax(a) + Bx(b) = 0.$$

The theorems for these problems take the same general form as Theorems V.28 and V.30 respectively, but with somewhat less concrete hypotheses. We turn now to the second order case.

We consider now the problem

$$x'' = f(t,x,x') \tag{V.29}$$
$$0 = g(x(a),x(b),x'(a),x'(b) \tag{V.30}$$

where
$$f:[a,b] \times R^n \times R^n \to R^n$$
and
$$g: R^n \times R^n \times R^n \times R^n \to R^{2n}$$

are continuous. Let

$$X = C'[a,b]$$

$$Z = C[a,b] \times R^{2n}$$

$$\text{dom } L = C^2[a,b]$$

$$L: \text{dom } L \to Z,\ x \to (x'',0)$$

$$N: X \to Z, x \to (f(\cdot,x(\cdot),\ x'(\cdot)),\ g(x(a),\ x(b),\ x'(a),\ x'(b))).$$

In this case we have

$$\text{Ker } L = \{x : x = c(t - a) + d, c,d \in R^n\}$$

$$\text{Im } L = C[a,b],\ \times\{0\}.$$

$$\dim \text{Ker } L = \text{codim Im} L = 2n.$$

Since Im L is closed, L is Fredholm of index 0. We take

$$Px = \frac{x(b) - x(a)}{b-a}(t-a) + x(a)$$

$$Qz = Q(y,\gamma) = (0,\gamma)$$

$$K_P z = K_P(y,p) = \int_a^b G(s,t)\ y(s)ds$$

where G(s,t) is the Green's function for the Picard problem.

Then

$$K_P(I-Q)\ Nx = \int_a^b G(s,t)\ f(s,x),\ x'(s)ds.$$

It is easily seen that N is L-compact on Ω for any open bounded $\Omega \subset X$.

Theorem V.34. Suppose

a) $\hat{G}_1$ is a convex autonomous curvature bound set relative to (V.29) such that $0 \in \hat{G}_1$ and for each $x_0 \in \partial \hat{G}_1$, $V_{1_{xx}}$ is positive semidefinite.

b) There exists a positive, continuous, nondecreasing function $\psi \in C^1[0,\infty)$ such that

$$\lim_{s \to +\infty} \frac{s^2}{\psi(s)} = +\infty$$

and

$$\|f(t,x,y)\| \leq \psi(\|y\|), \text{for } x \in \bar{\hat{G}}_1.$$

c) If $x(t)$ is a solution to

$$x'' = \lambda f(t,x,x') \qquad (\lambda \in (0,1)) \qquad \text{(V.31)}$$

$$0 = g(x(a), x(b), x'(a), x'(b)) \qquad \text{(V.32)}$$

with $x(t) \in \bar{\hat{G}}_1$ for $t \in [a,b]$ and $\|x'(t)\| \leq M$ for $t \in [a,b]$, where M is given by Lemma V.20, then $x(a), x(b) \notin \partial\hat{G}_1$.

d) If we define

$$\Omega_0 = \{(c,d) \in R^n \times R^n : \{c(b-a)+d \ , d\} \subset \hat{G}_1\},$$

then $g(d,c(b-a)+d, c,c) \neq 0$, for $(c,d) \in \partial\Omega_0$

and

$$d[g(c(b-a)+d,c,c),\Omega_0,0] \neq 0.$$

Then (V.29) - (V.30) has at least one solution.

Proof. By arguments similar to those in the proofs of Theorems V.19 and V.21 if we define

$$\Omega \equiv \{x \in X : x(t) \in \hat{G}_1, \|x'(t)\| \leq M_1 \equiv \max[M+1, \text{diam } \hat{G}_1/(b-a)], t \in [a,b]\}$$

then $x \in \partial\Omega$ and x being a solution to (V.31)-(V.32) implies that $x(a)$ or $x(b) \in \partial\hat{G}_1$. But this contradicts hypothesis c). Thus $x \notin \partial\Omega$.

We have

$$\text{Ker } L \cap \Omega = \{x : x = c(t-a) + d, \ c(t-a) + d \in \hat{G}_1, \ t \in [a,b], \ \|c\| < M_1\}.$$

By the convexity of $\hat{G}_1$ and the definition of M_1,

$$\text{Ker } L \cap \Omega = \{x : x = c(t-a) + d, \{c(b-a) + d, d\} \subset \hat{G}_1\}.$$

We also have

$$QN(c(t-a) + d) = g(d, c(b-a) + d, c, c).$$

We define $J : QZ = \{z : z = (0,(\alpha,\beta))\} \to \text{Ker } L$ by

$$Jz = \alpha(t-a) + \beta.$$

Then

$$d[JQN, \text{Ker } L \cap \Omega, 0] = d[g(d, c(b-a) + d, c, c), \Omega_0, 0] \neq 0.$$

Thus the Continuation Theorem yields existence.

Remark. Hypothesis a) in Theorem V.34 may be replaced by hypotheses a) and b) of Corollary V.22 or by hypothesis a) of Corollary V.23. As an example of more concrete hypotheses which can be used to replace c) and d) we give the following corollary.

Corollary V.35. *Suppose*

a) *There exists* R *such that if* $\|x\| \geq R$ *and* $x.y = 0$

$$\|y\|^2 + x.f(t,x,y) > 0$$

b) *There exists a positive, continuous, nondecreasing function* $\Psi \in C^1[0,\infty]$ *such that*

$$\lim_{s \to +\infty} \frac{s^2}{\psi(s)} = +\infty$$

and

$$\|f(t,x,y)\| \leq \psi(\|y\|) \quad (\text{for } \|x\| \leq R).$$

c)

$$\begin{bmatrix} x_1 \\ x_2 \end{bmatrix} \cdot g(x_1, x_2, y_1, y_2) > 0$$

for $\|x_1\| + \|x_2\| \geq R$ *and* $\|y_1\|, \|y_2\| \leq \max\,[M(R), 2R/(b-a)]$ (*where* M(R) *is the constant given by Lemma* V.20 *with* $\hat{G}_1 = \{x : \|x\| < R\}$).

<u>Then</u> (V.29)-(V.30) <u>has at least one solution.</u>

<u>Proof</u>. By the argument in the proof of corollary V.23

$$\hat{G}_1 = \{x : \|x\| - R < 0\}$$

satisfies hypothesis a) of Theorem V.34.

If x(t) is a solution to (V.31)-(V.32) with $\|x(t)\| \leq R$ on [a,b] and $\|x'(t)\| \leq M(R)$ on [a,b] and if $x(a) \in \partial G_1$ or $x(b) \in \partial G_1$, then

$$\|x(a)\| + \|x(b)\| \geq R$$

and

$$\begin{bmatrix} x(a) \\ x(b) \end{bmatrix} \cdot g(x(a),x(b),x'(a),x'(b)) > 0.$$

In particular, $g(x(a), x(b), x'(a),x'(b)) \neq 0$ which contradicts (V.32). Thus hypothesis c) of Theorem V.34 is satisfied.

In this case

$$\Omega_0 = \{(c,d) : \|c(b-a) + d\| < R, \|d\| < R\}.$$

We have

$$g(d,c(b-a)+d,c,c) \neq 0, \text{ for } (c,d) \in \partial\Omega_0$$

by the same argument as in the paragraph just preceding this one.

By the Poincare-Bohl theorem,

$$g(d,c(b-a) + d,c,c) \cdot \begin{bmatrix} d \\ c(b-a)+d \end{bmatrix} > 0, \quad (c,d) \in \partial\Omega_0$$

implies

$d[g(d,c(b-a)+d,c,c), \Omega_0, 0] = d[(d,c(b-a)+d),\Omega_0,0]$. The latter degree can be shown to be nonzero by direct computation.

We now give the boundary conditions a little more structure. Consider

$$x'' = f(t,x,x') \tag{V.33}$$

$$0 = g_1(x(a), x'(a)) \tag{V.34}$$

$$0 = g_2(x(b), x'(b))$$

where $f : [a,b] \times R \times R \to R$, $g_1 : R \times R \to R$, and $g_2 : R \times R \to R$ are continuous. Applying the techniques as above (and retaining the same operators specialized to $n = 1$) we obtain the following theorem and its generalization.

Theorem V.36. Suppose

a) There exists $R > 0$ such that

$$f(t,R,0) > 0, \quad f(t,-R,0) < 0.$$

b) $g_1(x,y)$ is nondecreasing in y with

$$g_1(R,0) < 0, \quad g_1(-R,0) > 0$$

c) $g_2(x,y)$ is nondecreasing in y with

$$g_2(R,0) > 0, \quad g_2(-R,0) < 0.$$

d) There exists $\psi(\sigma)$ continuously differentiable and positive on $[0,+\infty)$ such that

$$\int_0^\infty \frac{\sigma d\sigma}{\psi(\sigma)} = +\infty$$

and $|f(t,x,y)| \leq \psi(|y|)$ for $|x| \leq R$.

Then (V.33)-(V.34) has at least one solution.

Proof. We need to consider

$$x'' = \lambda f(t,x,x') \qquad (\lambda \in (0,1)) \tag{V.35}$$

$$0 = g_1(x(a),x'(a))$$
$$0 = g_2(x(b),x'(b)). \tag{V.36}$$

By exactly the same arguments that appeared in the proof of Theorem V.7, the set

$$G = \{(t,x,y) : |x| < R,\ |y| < h(|x|)\}$$

is a Nagumo set relative to (V.35) where $\sigma = h(\rho)$ is the unique solution to

$$\frac{d\sigma}{d\rho} = -\frac{\psi(\sigma)}{\sigma}$$

$$\sigma(-R) = N$$

and we choose N sufficiently large so that

$h(\rho) > 0$ on $[-R,R]$. For our purposes here we will take N large enough that

$$h(\rho) > 2R/(b-a),\ \rho \in [-R,R].$$

We define $\Omega \equiv \{x \in X : (t, x(t), x'(t)) \in G,\ t \in [a,b]\}$.

Suppose $x \in \partial\bar{\Omega}$ is a solution to (V.35)-(V.36). By Theorem V.6, $(a,x(a), x'(a)) \in \partial G$ or $(b, x(b), x'(b)) \in \partial G$. By arguments in the proof of Theorem V.7, $|x'(a)| < h(|x(a)|)$ and $|x'(b)| < h(|x(b)|)$, thus we must have $|x(a)| = R$ or $|x(b)| = R$. Suppose for definiteness that $x(a) = R$. Then $x'(a) \leq 0$ and we have

$$g_1(x(a), x'(a)) \leq g_1(R,0) < 0.$$

This contradicts (V.36) and we conclude that $x \notin \partial\bar{\Omega}$. Thus hypothesis a) of the Continuation Theorem is satisfied.

As in the proof of Theorem V.34 we have

$$\operatorname{Ker} L \cap \Omega = \{c(t-a) + d : (t, c(t-a)+d, c) \in G\}.$$

Since $|c(t-a) + d| < R$ implies that $|c| < 2R/(b-a) < h(|c(t-a) + d|)$, we have

$$\begin{aligned}\operatorname{Ker} L \cap \Omega &= \{c(t-a) + d : |c(t-a)+d| < R, t \in [a,b]\} \\ &= \{c(t-a)+d : |c(b-a)+d| < R,\ |d| < R\}.\end{aligned}$$

We also have

$$QN(c(t-a)+d) = \begin{bmatrix} g_1(d,c) \\ g_2(c(b-a)+d,c) \end{bmatrix}$$

We choose the isomorphism $J : \operatorname{Im} Q \to \operatorname{Ker} L$ defined by

$$J(c,d) = c(t-a) + d.$$

Then since $\{(t-a),1\}$ is a basis for Ker L,

$$d[JQN, \text{Ker } L \cap \Omega, 0] = d[(g_1(d,c), g_2(c(b-a)+d,c)), \Gamma, 0]$$

where

$$\Gamma = \{(c,d) : |c(b-a)+d| < R, \quad |d| < R\}.$$

The set Γ is tha parallelogram in (c,d) space bounded by the line segments

i) $c(b-a)+d = +R, \qquad -R \leq d \leq R$

ii) $c(b-a)+d = -R, \qquad -R \leq d \leq R$

iii) $d = R, \qquad \frac{-2R}{b-a} \leq c \leq 0$

iv) $d = -R, \qquad 0 \leq c \leq \frac{2R}{b-a}$

We observe that :

<u>On i)</u> We have $g_2(c(b-a)+d,c) = g_2(R,c) \geq g_2(R,0) > 0$

since

$$c = \frac{R-d}{b-a} \geq 0.$$

<u>On ii)</u>. We have

$$g_2(c(b-a)+d,c) = g_2(-R,c) \leq g_2(-R,0) < 0$$

since

$$c = \frac{-R-d}{b-a} \leq 0.$$

<u>On iii)</u>. We have

$$g_1(d,c) \leq g_1(R,0) < 0.$$

<u>On iv)</u>. We have

$$g_1(d,c) \geq g_1(-R,0) > 0.$$

Thus

$$d[(g_1(d,c), g_2(c(b-a)+d,c)), \Gamma, 0] \neq 0.$$

<u>Theorem V.37</u>. <u>Suppose</u>

a) <u>There exist</u> $\alpha, \beta \in C^2[a,b]$ <u>such that</u> $\alpha(t) < \beta(t)$ <u>on</u> $[a,b]$,

$$\alpha''(t) > f(t,\alpha(t),\alpha'(t)$$

<u>and</u> $(t \in [a,b])$

$$\beta''(t) < f(t,\beta(t),\beta'(t)).$$

b) $g_1(x,y)$ <u>is nondecreasing in</u> y <u>with</u>

$$g_1(\beta(a),\beta'(a)) < 0,\ g_1(\alpha(a),\alpha'(a)) > 0.$$

c) $g_2(x,y)$ <u>is nondecreasing in</u> y <u>with</u>

$$g_2(\beta(a),\beta'(a)) > 0,\ g_2(\alpha(a),\ \alpha'(a)) < 0.$$

d) <u>There exists</u> $\psi(\sigma)$, <u>continuously differentiable and positive on</u> $[0,+\infty)$ <u>such that</u>

$$\int_0^\infty \frac{\sigma\, d\sigma}{\psi(\sigma)} = +\infty$$

<u>and for</u> $|x| \leq \max[\max|\beta(t)|,\ \max|\alpha(t)|]$,

$$|f(t,x,y)| \leq \psi(|y|).$$

<u>Then</u> (V.33)-(V.34) <u>has at least one solution.</u>

<u>Proof</u>. Let N be choosen so that the unique solution to

$$\frac{d\sigma}{d\rho} = -\frac{\psi(\sigma)}{\sigma}$$

$$\sigma(-R_0) = N$$

is positive on $[-R_0, R_0]$ and so that $N > \max[\max_{[a,b]}|\alpha'(t)|,\ \max_{[a,b]}|\beta'(t)|]$.

Define

$$f^*(t,x,y,) = \begin{cases} f(t,x,N), & y > N \\ f(t,x,y), & |y| \leq N \\ f(t,x,-N), & y < -N \end{cases}$$

$$g_1^*(x,y) = \begin{cases} g_1(x,N), & y > N \\ g_1(x,y), & |y| \leq N \\ g_1(x,-N), & y < -N \end{cases}$$

$$g_2^*(x,y) = \begin{cases} g_2(x,N), & y > N \\ g_2(x,y), & |y| \leq N \\ g_2(x,-N), & y < -N. \end{cases}$$

Define

$$F(t,x,y) = \begin{cases} f^*(t,\beta(t),y) + (x - \beta(t)), & x > \beta(t) \\ f^*(t,x,y) \quad , & \alpha(t) \leq x \leq \beta(t) \\ f^*(t,\alpha(t),y) + (x - \alpha(t)), & x < \alpha(t). \end{cases}$$

$$G_1(x,y) = \begin{cases} g_1^*(\beta(a),y) - (x - \beta(a)), & x > \beta(a) \\ g_1^*(x,y), & \alpha(a) \leq x \leq \beta(a) \\ g_1^*(\alpha(a),y) - (x - \alpha(a)), & x < \alpha(a) \end{cases}$$

$$G_2(x,y) = \begin{cases} g_2^*(\beta(b),y) + (x - \beta(b)), & x > \beta(b) \\ g_2^*(x,y), & \alpha(b) \leq x \leq \beta(b) \\ g_2^*(\alpha(b),y) + (x - \alpha(b)), & x < \alpha(b). \end{cases}$$

We then make the following observation :

i) F, G_1, G_2 are continuous.

ii) $F(t,R,0) > 0$ and $F(t,-R,0) < 0$, for R sufficiently large.

iii) $G_1(x,y)$ is nondecreasing in y and

$$G_1(R,0) < 0, \quad G_1(-R,0) > 0$$

for R sufficiently large.

iv) $G_2(x,y)$ is nondecreasing in y and

$$G_2(R,0) > 0, \quad G_2(-R,0) < 0$$

for R sufficiently large.

v) For $|x| \leq R$, $\quad |F(t,x,y)| \leq M$.

Thus if R is choosen sufficiently large, Theorem V.36 applies to yield the existence of a solution $x_0(t)$ to

$$x'' = F(t,x,x'),\ G_1(x(a),\ x'(a)) = 0,\ G_2(x(b),\ x'(b)) = 0.$$

We now show that $\alpha(t) \leqslant x_0(t) \leqslant \beta(t)$. Suppose

$$\max_{[a,b]} [x_0(t) - \beta(t)] = x_0(t_0) - \beta(t_0) > 0.$$

If $t_0 = a$, then $x_0'(a) \leqslant \beta'(a)$, $\quad x_0(a) > \beta(a)$, and

$$G_1(x_0(a),\ x_0'(a)) \leqslant G_1(x_0(a),\ \beta'(a)) = g_1^*(\beta(a),\beta'(a)) - (x_0(a) - \beta(a))$$
$$< g_1^*(\beta(a),\beta'(a)) = g_1(\beta(a),\beta'(a)) < 0.$$

This contradicts (V.34). Similarly, $t_0 \neq b$.

If $a < t_0 < b$, then $x_0'(t_0) = \beta'(t_0)$ and

$$\beta''(t_0) \geqslant x_0''(t_0) = F(t_0,x_0(t_0),x_0'(t_0)) = f^*(t_0,\ \beta(t_0),\ \beta'(t_0)) +$$
$$(x_0(t_0) - \beta(t_0)) > f^*(t_0,\ \beta(t_0),\ \beta'(t_0)) = f(t_0,\ \beta(t_0),\ \beta'(t_0)).$$

which contradicts hypothesis a). Thus $x_0(t) \leqslant \beta(t)$.

By similar arguments $x_0(t) \geqslant \alpha(t)$. Thus $x_0(t)$ is a solution to

$$x'' = f^*(t,x,x'),\quad g_1(x(a),x'(a)) = 0,\quad g_2(x(b),x'(b)) = 0$$

But now

$$|f^*(t,x,y)| \leqslant \psi^*(|y|),\quad |x| \leqslant R_0,$$

where

$$\psi^*(\sigma) = \begin{cases} \psi(\sigma), & 0 \leqslant \sigma \leqslant N \\ \psi(N), & \sigma > N. \end{cases}$$

By earlier arguments (see the proof of Theorem V.8)

$$|x_0'(t)| \leqslant N.$$

Thus $x_0(t)$ is a solution to (V.33) - (V.34).

Example. In the theorems considered so far as applications of Theorem V.34, we have had $|d\,[g(d,c(b-a)+d,\ c,c),\Gamma,0]|$ equal to 1. To illustrate the potential of Theorem V.34 we conclude with an example where $|d[g(d,c(b-a)+d,c,c,\Gamma,0]| \neq 1$.

We consider the boundary value problem

$$x'' = f(t,x) \qquad (V.37)$$

$$[x(0)]^2 - [x'(0)]^2 = \alpha \qquad (V.38)$$

$$(\sqrt{\alpha^2+\beta^2} \neq 0)$$

$$2x'(1)\,[x(1) - x'(1)] = \beta \qquad (V.39)$$

where $f : [0,1] \times R \to R$ is continuous and $|f(t,x)| \leq M$ for $(t,x) \in [0,1] \times R$.

We will apply the Continuation Theorem to this problem using the operator formulation for problem (V.29) - (V.30).

Suppose $x(t)$ is a solution to

$$x'' = \lambda f(t,x)$$

satisfying (V.38) and (V.39) for some $\lambda \in (0,1)$. We have

$$|x(t)| \leq |x(0)| + |x'(0)| + M/2\ ,\ t \in [0,1] \qquad (V.40)$$

making use of $x(t) = x(0) + x'(0)t + \int_0^t \int_0^s x''(\tau)\, d\tau\, ds$.

We may write the boundary conditions in terms of $x(0)$ and $x'(0)$ as follows :

$$\alpha = [x(0)]^2 - [x'(0)]^2$$

$$\beta = 2x(0)x'(0) + 2x'(0)\left[\int_0^1 \int_0^s x''(\tau)\,d\tau\,ds - \int_0^1 x''(s)\,ds\right]$$

$$+ 2x(0)\int_0^1 x''(\tau)\,d\tau + 2\left[\int_0^1 x''(\tau)\,d\tau\right]\left[\int_0^1\int_0^s x''(\tau)\,d\tau\,ds - \int_0^1 x''(s)t\right].$$

If we let

$$\xi = x(0) + i\,x'(0),$$

then

$$\xi^2 = [x(0)]^2 - [x'(0)]^2 + 2\,x(0)\,x'(0)\,i$$

and the boundary conditions become

$$\alpha = \mathrm{Re}\,\xi^2$$

$$\beta = \mathrm{Im}\,\xi^2 + 2\,\mathrm{Im}\,\xi\,\left[\int_0^1 \int_0^s x''(\tau)\,d\tau\,ds - \int_0^1 x''(s)\,ds\right]$$

$$+ 2\,\mathrm{Re}\,\xi \int_0^1 x''(\tau)\,d\tau + 2\left[\int_0^1 x''(\tau)d\tau\right]\left[\int_0^1 x''(\tau)d\tau\,ds - \int_0^1 x'(s)ds\right]$$

Thus if $x(t)$ is a solution

$$|\xi|^2 = |\xi^2| = (\mathrm{Re}\,\xi^2)^2 + (\mathrm{Im}\,\xi^2)^2$$

$$\leq \alpha^2 + \beta^2 + K_1(M,\alpha,\beta)\,|\xi| + K_2(M).$$

This inequality is violated if $|\xi|$ is too large.

Thus

$$|x(0)|^2 + |x'(0)|^2 \leq K_3(M,\alpha,\beta).$$

Thus returning to (V.40), there exists $R_1(M,\alpha,\beta)$ such that

$$|x(t)| \leq R_1(M,\alpha,\beta).$$

Moreover,

$$|x'(t)| \leq |x'(0)| + \int_0^1 |x''(s)|\,ds \leq R_2(M,\alpha,\beta).$$

Thus, there exists R independent of λ and x such that

$$\|x(t)\|_1 \leq R.$$

If we define

$$\Omega \equiv \{x \in C^1[0,1] : \|x\|_1 = \max_{[0,1]} |x(t)| + \max_{[0,1]} |x'(t)| < R\}$$

the first hypothesis of the Continuation Theorem is satisfied.

We have

$$\text{Ker } L \cap \Omega = \{x(t) = c\,t + d : \max_{[0,1]} |c\,t + d| + |c| < R\}.$$

In terms of the basis $\{1,t\}$ for Ker L, Ker $L \cap \Omega$ may be expressed as

$$\tilde{\Omega} = \{(c,d) : \max_{[0,1]} |ct + d| + |c| < R\}$$

which is a neighborhood of the origin in R^2. In particular, if $(c,d) \in \partial\tilde{\Omega}$, we have $\max\{|c + d| + |c|, 2|c|\} = R \Rightarrow c^2 + d^2 > \frac{R^2}{16}$.

We may represent $JQN|\text{Ker } L$ in terms of $(1,t)$ as

$$\Gamma(c,d) = (d^2 - c^2 - \alpha, 2\,cd - \beta).$$

We may assume w. l . o. g that R is sufficiently large that

$$(c,d) \in \partial\tilde{\Omega}$$

implies that

$$(d^2 - c^2)^2 + (2cd)^2 = c^2 + d^2 > \alpha^2 + \beta^2.$$

Then $\Gamma(c,d) \neq 0$ for $(c,d) \in \partial\tilde{\Omega}$.

Moreover,

$$|d\,[JQN, \text{Ker } L \cap \Omega, 0]| = |d\,[\Gamma, \tilde{\Omega}, 0]| = 2.$$

Thus, the continuation theorem implies that (I.37) - (I. 38) - (I. 39) has a solution.

8. We begin this section by considering existence of periodic solutions of the Duffing equation

$$x'' + cx' + ax + bx^3 = p(t) \qquad (V.41)$$

We assume $p(t)$ is continuous on R, not identically 0, and T-periodic. We assume a,b and c are real constants. We consider three cases.

1) $b < 0$. In the notation of this chapter

$$f(t,x,x') = -cx' - ax - bx^3 + p(t).$$

We have

$$f(t,R,0) = -aR - bR^3 + p(t).$$

Since $-b > 0$, $f(t,R,0) > 0$ on $[0,T]$ for R sufficiently large. Similarly, $f(t,-R,0) < 0$ for R sufficiently large.

Moreover,

$$|f(t,x,x')| \leq |c| \ |x'| + D(R) = \psi(|x'|)$$

for $|x| \leq R$, and

$$\int_0^\infty p\,dp/\psi(p) = +\infty.$$

By Theorem V.7, (V.41) has at least one periodic solution.

2) $b > 0$, $a < 0$. Note that

$$\max_{x>0} [-bx^3 - ax] = \frac{-2a}{3}\left(\frac{-a}{3b}\right)^{\frac{1}{2}} \equiv M$$

$$\min_{x<0} [-bx^3 - ax] = \frac{2a}{3}\left(\frac{-a}{3b}\right)^{\frac{1}{2}} \equiv -M,$$

where the maximum and minimum occur at

$$\beta_1 = \left(-\frac{a}{3b}\right)^{\frac{1}{2}}$$

and

$$\alpha_1 = -\left(\frac{-a}{3b}\right)^{\frac{1}{2}}$$

respectively.

If we suppose that

$$\max \left| p(t) \right| < M, \tag{V.42}$$

then

$$f(t,\beta_1,\beta_1') = -a\beta_1 - b\beta_1^3 + p(t) = M + p(t) > 0$$

$$f(t,\alpha_1,\alpha_1') = -a\alpha_1 - b\alpha_1^3 + p(t) = -M + p(t) < 0.$$

Thus applying Theorem V.1 we conclude that (V.41) has a periodic solution such that

$$\alpha_1 \leq x(t) \leq \beta_1.$$

3) <u>$b > 0,\ a > 0,\ c > 0$.</u> In this case we apply Corollary V.33. To that end we convert (V.41) to a first order system. In particular we have

$$\left.\begin{aligned} x' &= y \\ y' &= -cy - ax - bx^3 + p(t) \end{aligned}\right\} = f(x,y). \tag{V.43}$$

Let

$$W(x,y) = \frac{1}{2}\left(ax^2 + y^2 + \frac{4c}{4+c^2}xy\right) + \frac{b}{4}x^4.$$

We have

$$\text{grad } W \, . \, f(x,y) = \begin{bmatrix} ax + \dfrac{2c}{4+c^2}y + bx^3 \\ \\ y + \dfrac{2c}{4+c^2}x \end{bmatrix} \bullet \begin{bmatrix} y \\ \\ -cy-ax-bx^3 + p(t) \end{bmatrix}$$

$$= 2c(4+c^2)^{-1}y^2 - cy^2 + yp(t) - 2c^2(4+c^2)^{-1}xy - 2ac(4+c^2)^{-1}x^2$$

$$- 2cb(4+c^2)^{-1}x^4 + 2c(4+c^2)\,xp(t)^{-1}$$

$$= p(t)\ [y + 2c(4 + c^2)^{-1}\ x] - 2c(4 + c^2)^{-1}\ [ax^2 + bx^4 + cxy + (1 + c^2/2\ )\ y^2]$$

There exists $R > 0$ such that

$$\text{grad } W.\ f(x,y) < 0$$

for $x^2 + y^2 \geqslant R^2$.

Moreover,

$$W(x,y) \to +\infty$$

as $x^2 + y^2 \to +\infty$. Thus $W(x,y)$ is a guiding function satisfying the hypotheses of Corollary V.33. Thus (V.43) has at least one periodic solution.

4) We conclude this section with a generalization of the assertion of 1) above. The technique employed illustrates the fact that sets Ω (or a priori bounds on solutions) may sometimes be obtained by other devices than the geometric techniques described in this chapter. Note however, that the technique below hinges on the special structure of the nonlinearity. We consider the problem

$$x'' + f(x') + h(t,x) = 0 \qquad \text{(V.44)}$$

$$x(0) = x(1),\ x'(0) = x'(1) \qquad \text{(V.45)}$$

where

i) $f : R \to R$ is continuous, $h(t,x) : [0,T]xR \to R$ is continuous,

ii) $f(0) = 0$,

iii) $x\ h(t,x) < 0$ for $|x| \geqslant R$.

Suppose $x(t)$ is a T-periodic solution to

$$x'' = -\lambda[f(x') + h(t,x)]$$

for some $\lambda \in (0,1)$. If $\max_{[0,T]} x(t) = x(to)$, then by periodicity $x'(t_o) = 0$ and $x''(t_o) \leq 0$. If $x(t_o) \geq R$, then

$$x''(t_o) = -\lambda\,[\,0 + h(t_o, x(t_o))]$$

and by iii), $x''(t_o) > 0$. Thus $x(t) \leq x(t_o) < R$ on $[0,T]$.

Similarly, $x(t) > -R$ on $[0,T]$.

Now

$$\int_0^T [x''(t)]^2\,dt = -\lambda\Big[\int_0^T f(x'(s))\,x''(s)\,ds + \int_0^T x''(s)\,h(s,x(s))ds\Big]$$

We have

$$\int_0^T f(x'(s))\,x''(s)ds = \int_{x'(0)}^{x'(T)} f(u)\,du = 0.$$

If $|x(t)| \leq R$, applying the Cauchy-Schwartz inequality to the second term we obtain

$$\|x''\|_2^2 \leq \lambda\,\|x'\|_2\,\Big(\int_0^T h^2(s,x(s))\,ds\Big)^{\frac{1}{2}} \leq \lambda C\,\|x''\|_2$$

where $C = T^{\frac{1}{2}} \max_{|x|\leq R} |h(s,x)|$. Thus $\|x''\|_2 \leq \lambda C$.

We then have

$$|x'(t)| \leq |x'(t_o)| + \Big|\int_{t_o}^{t} x''(s)\,ds\Big| \leq |x(t_o)| + |x''|_2\,T^{\frac{1}{2}}$$

Moreover, since $x(0) = x(T)$ there exists t_o such that $x'(t_o) = 0$. Thus

$$|x'(t)| \leq T^{1/2}\,\|x''\|_2 \leq T^{1/2}\,\lambda C \leq T^{1/2}\,C.$$

If we define

$$\Omega = \{x : |x(t)| < R,\ |x'(t)| < CT^{1/2} + 1,\ t \in [0,T]\}$$

then arguments similar to those used to prove Theorem V.7 lead to the conclusion that (V.44) has a T-periodic solution.

Bibliographical Notes.

The use of coincidence degree to prove existence theorems for boundary value problems of the type considered in this chapter originated with Mawhin (J. Differential Equations 16 (1974) 257-269). The unifying concept of curvature bound set is a generalization of the concept of bounding function introduced by Mawhin in the same paper.

Theorems V.7, V.8, and V.9 were first proved by Knobloch (Math. Z. 82 (1963) 177-197) and Schmitt (Math. Z. 98 (1967) 200-207) using different techniques. Theorem V.10 was first proved by Nagumo (Proc. Phys.-Math. Soc. Japan 24 (1942) 845-851). Theorem V.11 is due to Jackson (Advances in Math. 2 (1968), 307-363) using the Schauder fixed point theorem and essentially the same modification procedure used here. Theorems V.12, V.14, and V.15 are special cases of a theorem of Gaines (J. Differential Equations, 12 (1972), (291-312). Theorem V.15 is also a special case of a theorem in a much overlooked paper of Epheser (Math. Z. 61 (1955) 435-454). Theorem V.16 is adopted from a personal communication of W. Walter. Theorem V.17 is taken from Gaines (J. Differential Equations 16 (1974) 186-199).

Lemma V.20 is due to Schmitt (Proc. Streiermark. Math. Symposium, Graz, Austria, 1973). Another form of the Nagumo condition for systems is due to Hartman (Trans. Amer. Math. Soc. 96 (1960) 493-509). Corollary V.22 was first proved by Bebernes and Schmitt (J. Differential Equations 13 (1973) 32-47). Hypothesis a) of Corollary V.23 was first introduced by Hartman in the paper cited above where the Picard problem version of Corollary V.23 was proved (see Theorem V.24). Corollary V.23 itself was first proved by Knobloch (J. Differential Equations 9 (1971) 67-85) assuming f is locally Lipschitzian and using Cesari's method.

Schmitt (J. Differential Equations 11 (1972) 180-192) removed the Lipschitz assumption but imposed other auxilliary assumptions. The latter assumptions were then removed by Bebernes and Schmitt (J. Differential Equations 13 (1973) 32-47) using the operator of translation and Brouwer degree. Corollary V.26 is a weaker version of a result of Lasota and Yorke (J. Differential Equations 11 (1972) 509-518).

Corollary V.31 was first proved by Gustafson and Schmitt (Proc. Amer. Math. Soc. 42 (1974) 161-166). The guiding function concept and Corollary V.33 may be found on Krasnosel'skii (Russian Math. Surveys 21 (1966) 53-74).

Theorems V.36 and V.37 were first proved by Erbe (J. Differential Equations 7 (1970) 459-472) by a somewhat different technique. Example 3) of the last section is due to Reissig (Periodic Solutions-Forced Oscillations, Study # 6, Stability, Solid Mechanics Division, University of Waterloo). Example 4) can be found in more general form in Rouche and Mawhin ("Equations Différentielles Ordinaires", tome II ; Masson, 1973).

Papers closely related to the results and techniques of this chapter include the work of George and Sutton (Proc. Amer. Math. Soc. 25 (1970) 666-671), Bebernes (Proc. Amer. Math. Soc. 42 (1974) 121-127), Corduneanu (Acad. R.P. Roum., Fil. Iasi, Studi Cerc. Stunt. Mat. 8 (1957) 107-126 ; Rend. Acc. Sci. Fis. Mat. Napoli 25 (1958) 96-106), Bebernes, Gaines, and Schmitt (Ann. Soc. Sci. Bruxelles 88 (1974) 25-36), Mawhin and Muñoz (Ann. Mat. Pura Appl. (IV) 96 (1973) 1-19), and Fucik and Mawhin (Rapp. Sém. Appl. Méc. Univ. Louvain, 78 (1974).

There is a vast additional litterature concerning nonlinear boundary value problems including much important work not touched upon here. We refer the reader to the extensive bibliographies in the following : Bernfeld and Lakshmikantham ("An introduction to nonlinear boundary value problems", Academic Press, 1974), Bailey, Shampine, and Waltman ("Nonlinear two point boundary value problems", Academic Press, 1968), Mawhin ("International Conference on Differential Equations", H.A. Antosiewicz ed., Academic Press, New York, 1975, 537-556) and ("Difford 74, Summer School on Ordinary Differential Equations", Part I, J.E. Purkyne Univ., Brno, 1974, 37-60).

VI. APPROXIMATION OF SOLUTIONS - THE PROJECTION METHOD

1. In the preceding chapter we have employed the fundamental existence theorem of coincidence degree : *If the coincidence degree is defined and* $d[(L,N),\Omega] \neq 0$, *then* $Lx = Nx$ *has a solution in* $\Omega \cap \text{dom } L$. In this chapter we develop a theory of approximation for such solutions. This theory involves projection or Galerkin methods which provide natural approximation schemes to accompany our degree theoretic structure.

We will ultimately generate candidates for approximate solutions in the following manner : Let $\{X_m\}$ be a sequence of subspaces of $X \cap \text{dom } L$ such that $\dim X_m = m$. Let $\{Z_m\}$ be a sequence of subspaces of Z with $\dim Z_m = m$ and let R_m be a projector on Z such that $R_m Z = Z_m$. Define

$$\begin{aligned} L_m &: X_m \to Z_m,\ x_m \to R_m L x_m \\ N_m &: X_m \to Z_m,\ x_m \to R_m N x_m . \end{aligned} \qquad \text{(VI.1)}$$

We then consider the equation

$$L_m x_m = N_m x_m . \qquad \text{(VI.2)}$$

Under appropriate conditions (VI. 2) will have solutions which approximate solutions of

$$Lx = Nx. \qquad \text{(VI.3)}$$

More precisely, if we let

$$Sm = Sm(\Omega) = \{x_m : x_m \text{ is a solution to (VI. 2) in } X_m \cap \overline{\Omega}\},$$

$$S = S(\Omega) = \{x : x \text{ is a solution to (VI. 3) in } X \cap \overline{\Omega}\},$$

and

$$\rho(x_m, S) = \inf_{x \in S} \| x_m - x \|$$

we will seek conditions under which

$$\sup_{x_m \in S_m} \rho(x_m, S) \to 0.$$

We first consider a more general context where the pair (L_m, N_m) (not necessarily defined in terms of projections) converges to (L,N) in some sense and where $d[(L,N),\Omega] \neq 0$. We then study the special structure where L_m and N_m are defined by (VI. 1). Finally, we consider approximation of solutions when the operators arise from nonlinear boundary value problems of the type studied in chapter V.

2. We first give the fundamental theorem for this chapter which follows naturally from the definition of Leray-Schauder degree.

Theorem VI.1. Let $M_m : \bar{\Omega} \to X$ be compact for each positive integer m and let $M : \bar{\Omega} \to X$ be compact. Suppose

a) $\sup_{x \in \bar{\Omega}} \| Mx - M_m x \| \to 0$

b) $Mx \neq x,\ x \in \partial\Omega$

c) $d\,[I - M, \Omega, 0] \neq 0.$

If S_m denotes the solutions of $M_m x = x$ in $X_m \cap \bar{\Omega}$ and S denotes the solutions of $Mx = x$ in $\bar{\Omega}$, then $Sm \neq \phi$ for m sufficiently large, $S \neq \phi$ and

$$\sup_{x_m \in S_m} \rho(x_m, S) \to 0.$$

Proof. Let

$$0 < \mu = \inf_{x \in \partial\Omega} \|x - Mx\|.$$

(The existence of such a positive μ is a property of compact perturbations of the identity map). Let m be chosen sufficiently large so that

$$\|Mx - M_m x\| < \mu \text{ for } x \in \bar{\Omega}.$$

It follows from standard properties of Leray-Schauder degree that

$$d\,[I-M, \Omega, 0] = d[I-M_m, \Omega, 0] \neq 0.$$

Thus S_m and S are nonempty.

For $\varepsilon > 0$, let

$$S_\varepsilon = \bigcup_{x \in S} B(x,\varepsilon)$$

where $B(x,\varepsilon)$ is the ball of radius ε centered at x.
We want to show that for m sufficiently large, $S_m \subset S_\varepsilon$. Let

$$\mu_\varepsilon = \inf_{x \in \bar{\Omega} - S_\varepsilon} \|x - Mx\|.$$

Again we know that $\mu_\varepsilon > 0$. Choose m sufficiently large so that

$$\|Mx - M_m x\| \leqslant \mu_\varepsilon / 2$$

for $x \in \bar{\Omega}$. Then for $x \in \bar{\Omega} - S_\varepsilon$

$$\begin{aligned} \|x - M_m x\| &= \|x - Mx + Mx - M_m x\| \\ &\geqslant \|x - Mx\| - \|Mx - M_m x\| \\ &\geqslant \mu_\varepsilon - \mu_\varepsilon/2 = \mu_\varepsilon/2 > 0. \end{aligned}$$

Thus $S_m \subset S_\varepsilon$.

In order to give an analog of Theorem VI.1 in the context of coincidence degree we give first the following extension of Theorem III.3.

Lemma VI.1. Let L and L' be Frehhdm of index 0 and let L–L' have a continuous extension to all of X. Let N and N' be continuous on X and let $K_P(I - Q)$ be compact on Z. Suppose

a) $\inf\{\|Lx - Nx\| : x \in \bar{\Omega} \cap \mathrm{dom}L\} = \mu > 0.$

b) $\|(L - L')x\| + \|(N - N')x\| < \mu$ for $x \in \bar{\Omega} \cap \mathrm{dom}L.$

If the degree $d[(L'N'),\Omega]$ is defined, then $|d[(L',N'),\Omega]| = |d[(L,N),\Omega]|.$

Proof. We have

$|d[(L',N'),\Omega]| = |d[(L,L - L' + N')]|$

by the invariance of the absolute value of the coincidence degree. (This invariance was remarked at the beginning of chapter III. For a proof see Mawhin (A further invariance property of coincidence degree in convex spaces, Ann. Soc. Sci. Bruxelles Sér. I 87 (1973) 51 – 57).

We have for $x \in \bar{\Omega} \cap \mathrm{dom}L$

$\|(L - L' + N')x - Nx\| \leq \|(L - L')x\| + \|(N - N')x\| < \mu .$

By Theorem III.3, $d[(L,N),\Omega] = d[(L, L - L' + N'),\Omega]$.

Theorem VI.2. Let L and L_m be Fredholm of index 0. Let $K_P(I - Q)$ be compact and N and N_m be continuous. Suppose

a) $\sup\{\|L_m x - Lx\| : x \in \bar{\Omega} \cap \mathrm{dom}L\} \to 0.$

b) $\sup\{\|N_m x - Nx\| : x \in \bar{\Omega}\} \to 0.$

c) $Lx \neq Nx$, $x \in \partial\Omega$.

d) $d[(L,N),\Omega] \neq 0$.

If S_m denotes the solutions of $L_m x = N_m x$ in $\bar{\Omega}$ and S denotes the solutions of $Lx = Nx$ in $\bar{\Omega}$, then for m sufficiently large $S_m \neq \emptyset$ $S \neq \emptyset$ and

$$\sup_{x_m \in S_m} \rho(x_m,S) \to 0.$$

Proof. The proof is almost identical to the proof of Theorem VI.1 making use of Lemma III.2.

Remarks.

1) Theorem VI.2 doesn't require that L_m and N_m be defined in terms of projections as in (VI. 1).

2) Though the uniform convergence in hypothesis a) of Theorem VI. 2 seems rather natural, as we shall see later in this chapter, there are interesting approximation schemes in which $L_m x \to Lx$ pointwise but not uniformly. For this reason we give the following theorem.

Theorem VI.3. Let L be Fredholm of index 0, $\overline{\mathrm{dom}L} = X$, and N be L-compact on $\bar{\Omega}$. Let L_m be Freholm of index 0 and N_m be L-compact on $\bar{\Omega}$ with $\mathrm{Ker}L_m = \mathrm{Ker}L$ and $\mathrm{Im}\, L_m = \mathrm{Im}L$ for each m. Suppose

a) $L_m x \to Lx$ for each $x \in \mathrm{dom}L \cap \bar{\Omega}$.

b) There exists $c > 0$ such that

$$\|KL_m x\| \geq c\,\|x\|$$

for $x \in (I - P)\,\mathrm{dom}\,L$

c) $\sup\{\|N(x) - N_m(x)\| : x \in \bar{\Omega}\} \to 0$

d) $Lx \neq Nx$, $x \in \partial\Omega$

e) $d\,[(L,N),\Omega] \neq 0$.

Then the conclusion of Theorem VI. 1 holds.

Proof. We know that $Lx = Nx$ and $L_m x = N_m x$ are equivalent to

$$x = Px + JQNx + K_P(I - Q)Nx = Mx$$

$$x = Px + JQN_m + K_m(I - Q)N_m x = M_m x.$$

We know that M and M_m are compact on $\bar{\Omega}$. Note that for $z \in \mathrm{Im}L$,

$$K_m z = K_m L\, K_P z.$$

Using this fact we have

$$\|Mx - M_m x\| \leq \|JQ(N - N_m)x\| + \|(I - K_m L)K_P(I - Q)Nx\| + \|K_m(I - Q)\,(N_m - N)x\|.$$

From hypothesis c) the first term tends to o uniformly on $\bar{\Omega}$.
Hypothesis b) implies that

$$\|K_P L_m K_m L x\| \geq c \, \|K_m L x\|$$

or $\quad \frac{1}{c}\|x\| \geq \|K_m L x\|, \; x \in (I - P)\,\mathrm{dom}L.$

Moreover, for $z \in (I - Q)\, z = \mathrm{Im}L = \mathrm{Im}L_m$, we have (VI.4.)

$$\|K_m z\| = \|K_m L K_P\| \leq \frac{1}{c}\|K_P z\| \leq \frac{1}{c}\|K_P\|\|z\|$$

for all m, and the third term tends to 0 uniformly on $\bar{\Omega}$.

Let $x \in (I - P)\,\mathrm{dom}L$. We have

$$\begin{aligned} \|x - K_m L x\| &= \|x - K_m L_m x + K_m L_m x - K_m L x\| \\ &\leq \|x - x\| + \|K_m\| \, \|L_m x - Lx\|. \\ &\leq \frac{1}{c}\|K_p\| \, \|L_m x - Lx\| \end{aligned}$$

Thus $\quad \|x - K_m L x\| \to 0 \quad$ for each $x \in (I - P)\,\mathrm{dom}L$.

It follows from (VI. 4) that $K_m L$ can be *extended* to a bounded linear operator $\widetilde{K_m L}$ on $\overline{(I - P)\mathrm{dom}L} \equiv D_o$ in such a way that $\|\widetilde{K_m L}\| \leq \frac{1}{c}$. for all m. But then the mappings $\widetilde{K_m L}$ are equicontinuous on D_o and we must have

$$\|y - \widetilde{K_m L}y\| \to 0$$

as $m \to +\infty$ for $y \in D_o$. But then

$$(I - \widetilde{K_m L})K_P \;(I - Q)Nx \to 0$$

uniformly on $\bar{\Omega}$ since $\overline{K_P \;(I - Q)N(\bar{\Omega})} \subset D_o$ is compact and $(I - \widetilde{K_m L})$ is an equicontinuous sequence.

Thus we may apply Theorem VI.1 to obtain the desired conclusion.

Remarks

1) Consider

$$X_o = \bigcup_j \overline{K_P(Bj(o)) \cap \mathrm{Im}L)}$$

where $Bj(o)$ denotes the ball of radius j centered at o. Then X_o is a linear subspace of X. If X_o is barrelled (it is sufficient that X_o be of second category) then hypothesis b) of Theorem VI.3 may be replaced by

b') $\|L_m x\| \geq C\|x\|$, for all m, $x \in (I - P)\mathrm{dom}\, L$

and

b") $\|KL_m x\| \geq C_m\|x\|$, $C_m > 0$, $x \in (I - P)\mathrm{dom}\, L$.

In this case we have from b') directly that

$$\|K_m z\| \leq \frac{1}{C}\|z\|, \quad z \in (I - Q)\, z.$$

From b"), $K_m L$ has a bounded extension $K_m L$ to D_0 for each m. Moreover, if $x \in X_0$, then there exists a sequence $\{x_k\} \subset K_P(B_j(0))$ for some j such that

$$x_k \to x_0,$$

and

$$K_m L\, x_k \to K_m L\, x_0.$$

Moreover,

$$\|K_m L\, x_k\| \leq \|K_m L\, K_P z_k\| \leq \|K_m\|\, \|z_k\| \leq j/C$$

for all m. Thus

$$\|K_m L\, x_0\| \leq j/C$$

for all m. Thus $K_m L$ is pointwise bounded on X_0. By the Banach-Steinhaus theorem $\{K_m L\}$ is equicontinuous on X_0. The remainder of the proof is then as before, observing that

$$\overline{K(I-Q)N(\overline{\Omega})} \subset X_0.$$

2) It is possible to relax the requirement that Im Lm = ImL in such a way that the resulting version of Theorem VI.3 can be applied to operators L_m and N_m defined by (VI.1). However, rather than create a more cumbersome version of Theorem VI.3, in the next section we will develop an independant treatment of the operators (VI.1) exploiting their simple and convenient stucture directly.

3. We turn now to the approximations L_m and N_m defined by (VI.1) exploiting their special structure in order to apply Theorem VI.1.

<u>Lemma VI.2.</u> <u>If L_m and N_m are defined by</u> VI.1, <u>then L_m is Fredholm of index</u> 0 <u>and N_m is L_m-compact on $\overline{\Omega} \cap X_m$ for any open bounded Ω.</u>

<u>Proof.</u> Since $\dim X_m = \dim Z_m = m$ we have

$$\dim \operatorname{Im} L_m = m - \dim \operatorname{Ker} L_m.$$

Thus $\operatorname{codim} \operatorname{Im} L_m = \dim \operatorname{Ker} L_m$.

Since $\operatorname{Im} L_m$ is closed in Z_m; L_m is Fredholm of index 0. Moreover, N_m and K_m are continuous and have finite dimensional range and

$$\overline{K_m(I-Q_m)\, N_m(\overline{\Omega})}$$

is thus compact (where Q_m is the projector associated with K_m).

<u>Lemma VI.3.</u> <u>Let R_m be a projector of Z onto Z_m. Let L be Fredholm of index</u> 0 <u>with $\overline{\operatorname{dom} L} = X$. Suppose</u>

a) $Z_m \supset QZ$, $QR_m = R_m Q = Q$

b) $K_p (I - Q) R_m L$ is a continuous operator on domL.

Then if

$T_m \equiv P + K_p (I - Q) R_m L$:

i) The continuous extension $\tilde{T}_m$ of T_m to X is a projector on X,

ii) $R_m = Q + LT_m K_p (I - Q)$,

and

iii) If $T_m X \equiv X_m$, then $\mathrm{Ker} L \subset X_m$, $X_m \subset \mathrm{dom} L$, $\dim X_m = \dim Z_m$, $T_m P = PT_m$, and $LX_m = (I - Q) Z_m$.

Proof. We have the continuity of $\tilde{T}_m$ by hypothesis. Moreover, it is easily shown by direct computation that $T_m^2 = T_m$. Thus, for $x_0 \in X$

$$\tilde{T}_m^2 x_0 = \lim T_m^2 x_k = \lim T_m x_k = \tilde{T}_m x_0$$

and $\tilde{T}_m$ is a projector.

We have from a) that $QR_m = Q$ and

$$Q + L (P + K_p (I - Q)R_m L) K_p (I - Q) = Q + (I - Q)R_m (I - Q)$$

$$= Q + (R_m - Q) (I - Q)$$

$$= Q + R_m - Q - Q + Q$$

$$= R_m.$$

Finally, suppose $T_m x = x$. Then

$$Lx = LT_m x = (I - Q)R_m Lx \in (I - Q) Z_m.$$

Conversely, if $z \in (I - Q) Z_m$, then

$$T_m K_p z = (P + K_p (I - Q) R_m L) K_p z$$

$$= K_p z$$

thus $K_p z \in X_m$, and $LK_p z = z \in LX_m$. The other conclusions of iii) follow by direct calculation.

Remark. Conditions a) and b) are also necessary conditions for the conclusion of Lemma VI. 3.

Lemma VI. 4. Let T_m be a projector of X onto $X_m \subset \mathrm{dom} L$. Let L be Fredholm of index 0. Suppose $\mathrm{Ker} L \subset X_m$, $T_m P = PT_m$, and $LT_m K_p (I - Q)$ is continuous, then

i) $R_m = Q + LT_m K_p (I - Q)$ is a projector of Z onto $Z_m \equiv R_m Z$

ii) $T_m|\mathrm{dom}L = P + K_p(I - Q)R_mL$

and

iii) $LX_m = (I - Q)Z_m$.

Proof. Direct calculation.

Lemma VI.5. If L_m and N_m are defined by (VI. 1), $\mathrm{Ker}L \subset X_m$, $QR_m = R_mQ$, $LX_m = (I - Q)Z_m$, and $QZ \subset Z_m$, then

$$L_mx_m = N_mx_m \quad (x_m \in X_m)$$

is equivalent to

$$x = Px + JQNx + K_P(I - Q)R_mNx \equiv M_mx. \tag{VI.5}$$

If (X_m, T_m) and (Z_m, R_m) are related as in Lemma VI.3 (or Lemma VI.4), then $L_mx_m = N_mx_m$ is equivalent to

$$x = Px + JQNx + T_mK_p(I - Q)Nx \equiv M_mx. \tag{VI.6}$$

(or equivalently, $x = T_m Mx$)

Proof. Since L_m is Fredholm of index 0, $L_mx_m = N_mx_m$ is equivalent to

$$x_m = P_mx_m + J_mQ_mR_mN_mx_m + K_m(I - Q_m)R_mNx_m \tag{VI.7}$$

where P_m and Q_m are the projections associated with L_m and K_m is the associated generalized inverse. But we have

$$\mathrm{Ker}L_m = \mathrm{Ker}L$$

since $L_m x_m = R_mLx_m = Lx_m$, and $\mathrm{Ker}L \subset X_m$. Thus we may take $P_m = P|X_m$. Similarly

$$\mathrm{Im}\, L_m = (I - Q)Z_m$$

and we may take $Q_m = Q|Z_m$ (Note $QZ = Q_m Z_m$)

Moreover, we may take

$$K_m = K_p|Z_m.$$

Thus using $Q R_m = Q$ (VI.7) may be written as

$$x_m = P x_m + JQNx_m + K_p(I - Q) R_mN x_m.$$

The operator on the right may be extended to all of X with range contained in X_m. Thus any solution x to VI.5 is in X_m and is a solution to $L_mx = N_mx$. Moreover, any solution to $L_mx = N_mx$ is a solution to VI.5.

If we substitute

$$R_m = Q + LT_m K_p (I - Q)$$

we obtain

$$M_m x = Px + JQNx + K(I - Q)L\, T_m Kp(I - Q)Nx$$
$$= Px + JQNx + T_m K_p (I - Q)Nx$$

which yields the equivalent equation (VI. 6).

Theorem VI. 4. Let L be Fredholm of index 0 and N be L- compact on $\bar{\Omega}$. Let $\{(X_m, T_m)\}$ and $\{(Z_m, R_m)\}$ be sequences of subspaces and associated projectors which are related as in Lemmas VI. 3 and VI.4. Let L_m and N_m be defined by VI. 1. Suppose

a) $\|(I - T_m)\, x\| \leq \mu(m)\, \|Lx\|$, $\mu(m) \to 0$ as $m \to +\infty$ for $x \in \mathrm{dom} L$.

b) $Lx \neq Nx$, $x \in \partial\Omega$.

c) $d\,[\,(L,N), \Omega] \neq 0$.

Then $S_m \neq \emptyset$ for m sufficiently large, $S \neq \emptyset$ and

$$\sup_{x_m \in S_m} \rho(x_m, S) \to 0.$$

Proof. By Lemma VI. 5 $L_m x_m = N_m x_m$ is equivalent to

$$x = Px + JQNx + T_m K_p (I - Q)Nx = M_m x$$

while $Lx = Nx$ is equivalent to

$$x = Px + JQNx + K_p(I - Q)Nx \equiv Mx.$$

We have

$$\|Mx - M_m x\| = \|(I - T_m)\, K_p (I - Q)Nx\| \leq \mu(m)\, \|(I - Q)Nx\|.$$

For $x \in \bar{\Omega}$, $\|(I - Q)\, Nx\|$ is bounded. Thus

$$\sup_{x \in \bar{\Omega}} \| Mx - M_m x\| \to 0.$$

The conclusion of Theorem VI. 4 then follows from Theorem VI. 1 and the definition of coincidence degree.

Theorem VI. 5. The conclusion of Theorem VI. 4 remains valid if a) is replaced by

a') $\|(I - T_m)\, x\| \to 0$, $x \in X$ (or $\|(I - \tilde{T}_m)x\| \to 0$ in the content of Lemma VI.3)

Proof. Since $(I - T_m)$ is continuous for each m and $(I - T_m)\, x \to 0$ for each $x \in X$, by the Uniform Boundedness Principle, $\|I - T_m\| < B$ for some $B > 0$ and all m. In particular $\{I - T_m\}$ is an equicontinuous family. Since

$K_\rho(I - Q)N(\bar{\Omega})$ is relatively compact

$$\sup_{x \in \bar{\Omega}} \| (I - T_m)K_p(I - Q)N(x)\| \to 0.$$

Hence

$$\sup_{x \in \bar{\Omega}} \| Mx - M_m x\| \to 0$$

as in the proof of Theorem VI.4.

Theorem VI.6. Let L be Fredholm of index 0 and N be L- compact on $\bar{\Omega}$. Let L_m and N_m be defined by VI.1. where $LX_m = (I - Q)Z_m$, $\mathrm{Ker}L \subset X_m$, $QR_m = R_m Q$, $QZ \subset Z_m$. Suppose

a") $\|K_p(I - R_m)z\| \leq \mu(m)\|z\|$, $\mu(m) \to 0$, $z \in \mathrm{Im}L$.

b) $Lx \neq Nx$, $x \in \partial\Omega$.

c) $d[(L,N),\Omega] \neq 0$.

Then the conclusion of Theorem VI.4 remains valid.

Proof. Almost identical to the proof of Theorem VI.4 replacing (VI.6) by (VI.5).

Remarks.

1) It is possible to extend Theorem VI.6 to the case where LX_m is isomorphic to $(I - Q)Z_m$ with R_m being an isomorphism.

2) Suppose X is a separable Hilbert space and $\{\Phi_j\}_{j=1}^{\infty}$ is a complete orthonormal sequence in X such that $\{\Phi_j\} \subset \mathrm{dom}L$ and $\Phi_1, \Phi_2, \dots, \Phi_k$ is a basis for KerL. If we define

$$X_m = \mathrm{span} < \Phi_1, \Phi_2, \dots, \Phi_m >$$

and T_m is orthogonal projection onto X_m, then $\|(I - T_m)x\| \to 0$ for $x \in X$; i.e., a') is satisfied.

3) We have formulated (VI. 2) in such a way that explicit knowledge of K_P isn't required. If K_P is known, it is simpler to define the approximation procedure directly in terms of the equation.

$$x = Px + JQNx + K_P(I - Q)Nx = Mx.$$

In this case the approximation would be simply

$$x_m = T_m M x_m \equiv M_m x_m \quad (x_m \in X_m) \qquad \text{(VI.2)}^*$$

where T_m is a projector of X onto a finite dimensional subspace $x_m \subset X$.

Since M is compact, it is again sufficient to have $\|(I - T_m)x\| \to 0$ for $x \in X$. If X is a separable Hilbert space we obtain a particulary simple form of the approximation theorem. Let $S_m(\Omega)^*$ denote the set of solutions to $(VI.2)^*$ in Ω.

Theorem VI.7. Let L be Fredholm of index 0 and let N be L-compact on $\overline{\Omega}$. Let X be a separable Hilbert space. Suppose

a) $Lx \neq Nx$, $x \in \partial\Omega$

b) $d[(L,N),\Omega] \neq \phi$

If $\{\Phi j\}_{j=1}^{\infty}$ is a complete orthonormal sequence in X and T_m is defined to be orthogonal projection of X onto span $< \Phi_1, \Phi_2, \ldots, \Phi_m >$, then $S_m^* \neq 0$ for m sufficiently large $S \neq \phi$ and sup $\rho(x_m, S) \to 0$.

$$x_m \in S_m^*$$

Proof. We have

$$\|(I - T_m)x\| \to 0.$$

for each $x \in X$. The proof is then identical to the proof of Theorem VI.5.

3. We now apply the approximation theory of the preceding section to the case where L is a differential operator. In particular, we will show how to obtain approximation theorems to accompany the existence theorems of Chapter V. To simplify our exposition we will confine our attention to operators defined by first order systems of equations. The details of modifications necessary to adapt the techniques to second order systems or nth order scalar equations are rather straightforward and are left for the reader.

We consider the general problem

$$x' = A(t)x + f(t,x)$$
$$Cx(a) + Dx(b) = 0$$

where A(t) is a continuous n x n matrix on [a,b], $f : [a,b] \times R^n \to R^n$ is continuous and C and D are constant n x n matrices. We define L_o and N_o as follows :

$$X = C[a,b] \cap \{x : Cx(a) + Dx(b) = 0\}$$
$$Z_o = L_2 [a,b]$$
$$\text{dom } L_o = H [a,b](\{x : x \text{ is absolutely continuous on } [a,b], x' \in L^2 [a]) \cap X$$

$$L_o : \text{dom } L_o \to Z_o, \; x \mapsto x' - A(t)x$$
$$N_o : X \to Z_o, \; x \mapsto f(.,x(.)).$$

The usual norm on L_2 [a,b] will be denoted by $\|\cdot\|_2$.
The space H[a,b] is a Hilbert space under the inner product [f,g] = (f,g) + (f',g'). where $(f,g) = \int_a^b f(t).g(t)dt$. Note that in Chapter V when considering first order systems we defined L and N using

$$X = C\,[a,b] \cap \{x : Cx(a) + Dx(b) = 0\}$$

$$Z = C\,[a,b]$$

$$\mathrm{dom}L = C^1\,[a,b] \cap X.$$

$$L : \mathrm{dom}L \to Z,\ x \mapsto x' - A\,(t)x$$

$$N : X \to Z,\ x \mapsto f(.,x(.)).$$

We will show that these formulations are equivalent from the point of view of coincidence degree. For that and other purposes we will make use of the adjoint operator L_o^* of L_o. This operator is defined by :

$$\mathrm{dom}\ L_o^* = H[a,b] \cap \{x : C^*x(a) + D^*x(b) = 0\}$$

$$L_o^* : \mathrm{dom}\ L_o^* \to Z_o,\ x \mapsto -x' - A^T(t)x$$

where C^* and D^* define a set of adjoint boundary conditions. We will assume throughout that $[C,D]$ has rank n which implies that $[C^*, D^*]$ has rank n. For detailed discussion of the definition and properties of the adjoint boundary conditions we refer the reader to Cole ("Theory of Ordinary Differential Equations", Appleton, Century, Crofts, New York, 1968). We will make use of the following properties without proof :

1) $\mathrm{Im}L_o = (\mathrm{Ker}\ L_o^*)^{\perp}$

2) $\dim \mathrm{Ker}L_o = \dim \mathrm{Ker}L_o^*$.

The kernels of L_o and L_o^* are, of course, finite dimensional. These two statements yield almost immediately.

<u>Lemma VI.6</u>. <u>If</u> rank $[C,D] = n$, <u>then</u> L_o <u>is Fredholm of index</u> 0.

<u>Proof</u>. We have from 1) and 2) above

$$\dim \mathrm{Ker}\ L_o = \mathrm{codim}\ \mathrm{Im}\ L_o.$$

If $\{z_k\} \subset \mathrm{Im}\ L_o$ and $z_k \to z_o \in L_2[a,b]$, then

$$\int_a^b z_k(s).y(s)ds \to \int_a^b z_0(s).y(s)ds = 0$$

for each $y \in \text{Ker } L_o^*$. Thus $z_0 \in (\text{Ker} L_o^*)^{\perp} = \text{Im } L_o$.
Thus Im L_o is closed.

<u>Lemma VI.7.</u> $d\,[(L_o,N_o),\Omega]$ <u>is defined if and only if</u> $d[L,N),\Omega]$ <u>is defined. Moreover,</u>

$$|d\,[(L,N),\Omega] = |d\,[(L_o,N_o),\Omega]|.$$

<u>Proof.</u> Suppose $d\,[(L_o,N_o),\Omega]$ is defined. We have Ker $L \subset$ Ker Lo. Suppose $x \in$ Ker Lo. Then

$$x'(t) = A(t)x(t)$$

and the continuity of $x(t)$ implies the continuity of $x'(t)$. Thus $x \in$ Ker L. Thus Ker L = Ker L_o. We may take $P_o \equiv P$.

Let Q_o be the orthogonal projector of Z_o onto Ker L_o^*. We have Im L $\subset$ Im L_o, thus $(I - Q_o)|'Z = (I - Q_o)|$ $[a,b] \supset$ Im L. On the other hand if $z \in (I - Q_o)|Z$, then

$$z = (I - Q_o)|z_o = z_o - Q_o z_o$$

where $z_o \in Z$. Since z_o is continuous and $Q z_o \in \text{Ker } L_o^*$ is continuous, z is continuous.

Moreover, for some $Lx = z \in \text{dom } L_o$
or $x'(t) = A(t)x(t) + z(t)$. But the continuity of $A(t)$, $x(t)$ and $z(t)$ imply the continuity of $x'(t)$. Thus $z \in$ Im L. Thus $(I - Q_o)|Z$ = Im L.

We know that $Q \equiv Q_o|Z$ has the property that $Q^2 = Q$ and Q is linear. Moreover QZ = Ker L_o^*. Since Ker L_o^* is finite dimensional, the $L_2\,[a,b]$ and $C\,[a,b]$ norms induce équivalent norms $\|\cdot\|_2$ and $\|\cdot\|$ on Ker L_o^*. Thus for $z \in Z$

$$\begin{aligned}\|Qz\| &\leq \gamma_1\|Qz\|_2 \leq \gamma_1\|Q\|_2\,\|z\|_2 \\ &\leq \gamma\|Q\|_2\,(b-a)^{\frac{1}{2}}\,\|z\|.\end{aligned}$$

Thus Q is continuous on Z. It follows that L is Fredholm of index 0. It is easily seen that $J_o \equiv J$ and $K_P = K_{P_o}|Z$.

Note that $N_o = N$ and $N_o x \in Z$ for $x \in X$. Thus

$$K_P(I - Q)N = K_{P_o}(I - Q_o)N_o,$$

and N_o being L_o-compact implies N is L-compact. We have that $L_o x = N_o x$ is equivalent to

$$x = P_o x + J_o Q_o N_o x + K_{P_o}(I - Q_o)N_o x = M_o x$$

and $Lx = Nx$ is equivalent to

$$x = P_o x + J_o(Q_o|Z)Nx + (K_{P_o}|Z)(I - Q_o)Nx = Mx.$$

Since $N_o x = Nx \in Z$, $M_o = M$. By the definition of coincidence degree, the Lemma follows.

For our purposes it is usefull to have a more explicit representation for the projectors P_o and Q_o. Let $\Phi(t)$ be a matrix whose columns are a basis for Ker L_o. Suppose dim Ker $L_o = k$. Then $\Phi(t)$ is n x k. Let $\Psi(t)$ be a matrix whose columns are a basis for Ker L_o^*. We assume w.l.o.g. that the columns are orthonormal in $L_2[a,b]$. Thus we may define P_o and Q_o by

$$P_o x = \Phi(t)\int_a^b \Phi^T(s)x(s)ds \tag{VI.8}$$

$$Q_o z = \Psi(t)\int_a^b \Psi^T(s)z(s)ds \tag{VI.9}$$

Lemma VI.8. If P_o and Q_o are defined by (VI.8.) and (VI.9.), then :

a) P_o is a projector of X onto Ker L_o

b) Q_o is a projector of Z_o onto Ker L_o^*.

Proof. Straightforward verification.

Lemma VI.9. There exists an n x n matrix function h(s,t) defined on [a,b] x [a,b] such that :

a) h(s,t) is continuous except on $\{(t,s) : t = s\}$

b) $\|h(s,t)\|$ is bounded on [a,b] x [a,b]

c)
$$K_{P_o} z = \int_a^b h(s,t)z(s)ds \text{ for } z \in \text{Im } L_o.$$

<u>Proof.</u> Note that $K_{P_o} z$ is the unique solution to

$$L_o x = z$$
$$P_o x = 0.$$

The general solution to $x' = A(t)x + z(t)$ is given by

$$x(t) = F(t)d + F(t)\int_a^t F^{-1}(s)z(s)ds \qquad \text{(VI.10)}$$

where $F(t)$ is the principal fundamental matrix solution to $x' = A(t)x$. In order for $x(t)$ given by (VI.10) to satisfy $L_o x = z$ we must have

$$0 = Cx(a) + Dx(b) = Cd + D[F(b)d + F(b)\int_a^b F^{-1}(s)z(s)ds]$$

Thus d must satisfy

$$[C + DF(b)]d = - DF(b)\int_a^b F^{-1}(s)z(ds) \qquad \text{(VI.11)}$$

Let

$$B = C + DF(b).$$

We will make use of the following facts :

i) $\dim \operatorname{Ker} L_o = n - \operatorname{rank} B$

ii) There exists an invertible $n \times n$ matrix $B^\dagger$ such that for $c \in \operatorname{Range} B$, $d = B^\dagger c$ is the unique solution to $Bd = c$ such that $d \in \operatorname{Range} B^T$ (or equivalently ; $B_1^T d = 0$; where B_1 is an $n \times k$ matrix whose columns are a basis for Ker B).

iii) If $c \in \operatorname{Range} B$, then d is a solution to $Bd = c$ if and only if

$$d = B_1\alpha + B^\dagger c$$

where $\alpha \in R^k$.

iv) $c \in \operatorname{Range} B$ if and only if

$$B_2^T c = 0$$

where the columns of B_2 are a basis for $\operatorname{Ker} B^T$.

v) The columns of $F(t)B_1$ are a basis for $\operatorname{Ker} L_o$.

vi) The rows of $- B_2^T DF(b)F^{-1}(t)$ are a basis for Ker L_o^*. [This follows from the fact that $(F^{-1}(t))^T$ is the principal fundamental matrix solution of the adjoint system

$$- x' - A^T(t)\, x = 0$$

and from the fact that the adjoint boundary conditions

$$C^* x(a) + D^* x(b) = 0$$

are equivalent to

$$x(a) = C^T d$$

$$x(b) = - D^T d$$

for some $d \in R^n$. Thus

$$(F^{-1}(t))^T c$$

is in Ker L_o^* iff

$$(F^{-1}(a))^T c = C^T d$$

and

$$(F^{-1}(b))^T c = - D^T d,$$

or

$$c = F^T(a) C^T d$$

and

$$c = - F^T(b)\, D^T d.$$

There exists d satisfying these equations if and only if

$$0 = (C^T + F^T(b) D^T)\, d$$

or

$$0 = B^T d.$$

Thus the members of Ker L_o^* are given by

$$- (F^{-1}(t))^T F^T(b)\, D^T d$$

where $d \in$ Ker B^T. It follows that the columns of $- (F^{-1}(t))^T F^T(b) O^T B_2$ are a basis for Ker L_o^*.]

Returning to (VI.11) we have from iii) that d is a solution if and only if

$$d = B_1 \alpha - B^{\dagger} DF(b) \int_a^b F^{-1}(s)\, z(s) ds \qquad \text{(VI.12)}$$

and from (iv)

$$- B_2^T\, DF(b) \int_a^b F^{-1}(s) z(s) ds = 0.$$

From v) the latter requirement says that $z(t)$ must be orthogonal to the elements of Ker L_o^*; i.e., it is a requirement that $z \in \text{Im } L_o$.

Substituting back into (VI.10) we have the solutions to $L_o x = z$ represented by

$$x(t) = F(t)\, B_1\alpha - F(t)B^{\dagger}DF(b)\int_a^b F^{-1}(s)z(s)ds$$
$$+ F(t)\int_a^t F^{-1}(s)z(s)ds.$$

Imposing the condition $P_o x = 0$ we must have

$$0 = F(t)\, B_1\alpha - \Phi(t)\int_a^b \Phi^T(s)\, F(s)\, B^{\dagger}DF(b)ds\int_a^b F^{-1}(\tau)z(\tau)d\tau$$
$$+ \Phi(t)\int_o^b \Phi^T(s)F(s)\left(\int_o^s F^{-1}(\tau)z(\tau)d\tau\right)ds.$$

Since the columns of $F(t)B_1$ and the columns of $\Phi(t)$ are a basis for Ker L_o,

$$F(t)B_1 = \Phi(t)W$$

where W is $k \times k$ and invertible.

But then we must have

$$W\alpha = \left(\int_a^b \Phi^T(s)F(s)ds\right) B^{\dagger}DF(b)\left(\int_a^b F^{-1}(\tau)z(\tau)d\tau\right)$$
$$- \int_a^b \Phi^T(s)F(s)\left(\int_a^s F^{-1}(\tau)z(\tau)d\tau\right)ds.$$

Thus the solution we seek is

$$x(t) = \Phi(t)\left(\int_a^b \Phi^T(s)F(s)ds\right) B^{\dagger}DF(b)\left(\int_a^b F^{-1}(\tau)z(\tau)d\tau\right)$$
$$- \Phi(t)\int_a^b \Phi^T(s)F(s)\left(\int_a^s F^{-1}(\tau)z(\tau)d\tau\right)ds$$
$$- F(t)\, B^{\dagger}DF(b)\left(\int_a^b F^{-1}(\tau)z(\tau)d\tau\right)$$
$$+ F(t)\int_a^t F^{-1}(\tau)z(\tau)d\tau$$

$$= \int_a^b h(\tau,t)z(\tau)d\tau$$

where

$$h(\tau,t) = \begin{cases} \left[\Phi(t)\int_a^b \Phi^T(s)F(s)ds - F(t)\right] B^{\dagger}DF(b)F^{-1}(\tau) \\ \quad + \left[F(t) - \Phi(t)\int_\tau^b \Phi^T(s)F(s)ds\right] F^{-1}(\tau), & a \leq \tau \leq t \\ \left[\Phi(t)\int_a^b \Phi^T(s)F(s)ds - F(t)\right] B^{\dagger}DF(b)F^{-1}(\tau) & t < \tau \leq b \\ \quad + \Phi(t)\left(\int_\tau^b \Phi^T(s)F(s)ds\right) F^{-1}(\tau). \end{cases}$$

Corollary VI.1. N_o is L_o-compact on $\overline{\Omega}$ for any bounded open subset Ω of X.

Proof. Since f is continuous N_o is continuous from X into C[a,b] under the uniform norm. Hence N_o is continuous from X into $Z_o = L_2\,[a,b]$.

The compactness of $K_{P_0}\,(I - Q_o)\,N_o\,(\overline{\Omega})$ follows from the boundedness of Ω, the continuity of N_o and the boundedness of $h(\tau,t)$ using the Ascoli-Arzela theorem and standard arguments.

We are now prepared to define appropriate subspaces X_n of X and corresponding subspaces Z_n of Z_o. Let $\{\varphi_j\}_{j=R+1}^{\infty} \subset (I - P_o)\text{ dom } L_o$ be a linearly independent sequence whose span is dense in $(I - P_o)\text{ dom } L_o$ as a subspace of H[a,b].

Define $$X_m = \langle \varphi_{k+1}, \varphi_{k+2},\ldots, \varphi_m\rangle \oplus \text{ Ker } L_o. \qquad \text{(VI.13)}$$

Define $$Z_m = \langle L_o\varphi_{k+1}, L_o\varphi_{k+2},\ldots, L\varphi_m\rangle \oplus \text{ Ker } L_o^{*}.$$

Lemma VI.10. a) The sequence $\{L_o \varphi_i\}_{i=k+1}^{\infty}$ is linearly independent in $L_2[a,b]$.

b) The span of $\{L_o \varphi_i\}_{i=k+1}^{\infty}$ is dense in Im L_o.

Proof. a) The independence follows immediately from the fact that L_o is 1 - 1 on $(I - P_o)X$.

b) If $z \in \text{Im } L_o$ then $L_o x = z$ for $x = K_{Po} z \in (I - P_o) \text{ dom } L_o$. There exists a sequence $\{x_j\} \subset \langle \varphi_{k+1}, \varphi_{k+2}, \ldots \rangle$ such that $x_j \to x$ in $H[a,b]$. But then $L_o x_j \to L_o x$ since L_o is continuous from the $H[a,b]$ topology into $L_2[a,b]$. But

$$L_o x_j = L_o\left(\sum_{i=k+1}^{m_j} \alpha_{ji}\varphi_i\right) = \sum_{i=k+1}^{m_j} \alpha_{ji} L_o\varphi_i .$$

Thus $L_o x_j \in \langle L\varphi_{k+1}, L\varphi_{k+2}, \ldots \rangle$ and this completes the proof.

Let $\{\psi_1,\psi_2,\ldots\psi_k\}$ be an orthonormal basis for Ker $L_o^* = Q_o Z$. Then $\{\psi_1,\psi_2,\ldots\psi_k, L\varphi_{k+1}, L\varphi_{k+2},\ldots\}$ is a linearly independent sequence whose span is dense in $L_2[a,b]$. By the Gram-Schmidt process we can construct a complete orthonormal sequence $\{\psi_i\}_{i=1}^{\infty}$ in $L_2[a,b]$ such that $Z_m = \langle \psi_1,\psi_2,\ldots\psi_m \rangle$. Let R_m be orthogonal projection onto Z_m i.e.,

$$R_m z = \sum_{i=1}^{m} (\psi_i, z)\, \psi_i(t) \tag{VI.14}$$

Lemma VI.11. For $z \in \text{Im } L_o$

$$\|K_{P_o}(I - R_m)z\| \leq \mu(m)\|z\|_Z$$

where $\mu(m) \to 0$ as $m \to +\infty$.

Proof.

$$\|K_{P_o}(I - R_m)z\| = \left\| \int_a^b h(s,o) \sum_{i=m+1}^{\infty} (\psi_i, z)\psi_i(s)ds \right\|.$$

Since the rows of $h(s,t)$ are in $L_2[a,b]$ for fixed t we have

$$\| Kp_0(I - R_m)z\| = \| \sum_{i=m+1}^{\infty} (\psi_i,z) \int_a^b h(s,t)\ \psi_i(s)ds\|$$

$$\leq \sum_{j=1}^{n} \max_{[a,b]} \sum_{i=m+1}^{\infty} \left| \int_a^b h_j(s,t)\ \psi_i(s)ds \right| \ |(z,\psi_i)|$$

where h_j denotes the jth row of $h(s,t)$. Thus

$$\| Kp_0(I - R_m)z\| \leq \left(\sum_{i=m+1}^{\infty} (z,\psi_i)^2 \right)^{\frac{1}{2}} \sum_{j=1}^{n} \max \left[\sum_{i=m+1}^{\infty} \left(\int_a^b h_j(s,t)\ \psi_i(s)ds \right)^2 \right]^{\frac{1}{2}}$$

For each t the sequence

$$\sigma_m^j(t) = \left[\sum_{i=m+1}^{\infty} \left(\int_a^b h_j(s,t)\ \psi_i(s)ds \right)^2 \right]^{\frac{1}{2}}$$

converges to 0. Note that

$$\sigma_m^j(t) = \left[\| h_j(.,t)\|_2^2 - \sum_{i=1}^{m} \left(\int_a^b h_j(s,t)\ \psi_i(s)ds \right)^2 \right]^{\frac{1}{2}}$$

and the terms are the right are continuous as functions of t on $[a,b]$. Thus $\sigma_m^j(t)$ is continuous for each m. Moreover, $\{\sigma_m^j(t)\}$ is decreasing as m increases. By Dini's theorem the sequence $\{\sigma_m^j(t)\}$ converges to 0 uniformly. Thus

$$\| Kp_0(I - R_m)z\| \leq \|z\|_2 \sum_{j=1}^{n} \mu_j(m)$$

$$\leq \|z\|_2\ \mu(m)$$

where $\mu(m) \to 0$ as $m \to +\infty$.

Theorem VI.8. Let $\{X_m\}$ and $\{R_m\}$ be defined by (VI.13) and (VI.14). Suppose rank $[C,D] = n$. Suppose

a) $L_0 x \neq N_0 x$, $x \in \partial\Omega$.

b) $d[(L_0,N_0),\Omega] \neq 0$.

Then $S_m \neq \phi$ for m sufficiently large, $S \neq \phi$ and

$$\sup_{x_m \in S_m} \rho(x_m,S) \to 0$$

Proof. Apply Theorem VI.6 using Lemmas XI.6, and VI.11 and Corollary VI.1.

Remarks.

1) The construction above of R_m and X_m requires knowledge of Ker L and Ker L*. It may be possible to remove this requirement using the techniques of Locker (Trans. Amer. Math. Soc. 203 (1975) 175 - 183).

2) Construction of sequences $\{\varphi_j\}_{j=k+1}^{\infty}$ of the type needed in (VI.13) has been studied by Locker (Trans. Amer. Math. Soc. 154 (1971) 57 - 68). In the next paragraph we discuss a procedure which requires only construction of complete sequences in L_z [a,b].

3) Projections may also be defined in terms of finite difference approximation.

4) Though we have succeeded in approximating the original problem Lx = Nx by a *finite dimensional* problem $L_m x_m = N_m x_m$ we have not yet considered the problem of solving of the finite dimensional equation. Note that we know that

$$d\,[\,(L_m,N_m),\Omega \cap X_m] \neq 0$$

from the arguments given above. Equivalently, we know that

$$d\,[I - M_m,\Omega \cap X_m,0] \neq 0 \qquad \text{(VI.15)}$$

where M_m is given by (VI.6). The latter degree is an ordinary Brouwer degree. At the time these notes are being prepared much research is being devoted to the development of algorithms for approximating fixed points of finite dimensional maps. It appears likely that in the near future algorithms will exist which approximate solutions to $M_m x_m = x_m$ assuming only that the Brouwer degree of the mapping with respect to an appropriate set is nonzero. (At least, one can hope for schemes which are constructive in nature, if not computationally efficient). The current state of the subject is such that algorithms exist to approximate solutions in the case where M_m maps a cube (or other geometrically convenient set) into itself. We discuss a context in which this algorithm may be applied in the next paragraph.

5) Using the procedure developed in this section, one may obtain an approximation theorem to accompany each of the existence theorem of Part V. This will be illustrated in section 6.

5. We now show how to apply the simpler structure of Theorem VI. 7 to the boundary value problem of the preceding section. We reformulate the boundary value problem with

$$X_1 = L_2\,[a,b] \cap \{x : Cx(a) + Dx(b) = 0\}$$

$$Z_1 = L_2\,[a,b]$$

$$\text{dom } L_1 = H[a,b] \cap X_1$$

$$L_1 : \text{dom } L_1 \to Z_1,\; x \mapsto x' - A(t)\,x$$

$$N_1 : X_1 \to Z_1,\; x \mapsto f(.,x(.)).$$

It must be observed that continuity of f is not enough to guarantee that N_1 is defined as a map from $L_2\,[a,b]$ into $L_2\,[a,b]$. For this purpose we make the additional assumption : There exists $B > 0$ such that

$$\|f(t,x)\| \leq B \qquad \text{(VI.16)}$$

for $(t,x) \in [a,b] \times R^n$. We note that weaker assumptions than (VI.16) would be sufficient. However, *boundedness* is a simple assumption, and; moreover, under many hypotheses which imply existence of solutions to the boundary value problem, the nonlinear function f may be replaced by a bounded $\hat{f}$ which yields an equivalent problem. This device was used, for example, in the proof of Theorem V.8.

<u>Lemma VI.12.</u> <u>If</u> f <u>is bounded and continuous, then</u> N_1 <u>is continuous</u>.

<u>Proof.</u> Suppose $\{x_m\} \subset L_2\,[a,b]$ and $x_m \to x_o$ where the convergence is in L_2. Consider

$$g_m(t) = f(t,x_m(t)).$$

Since f is continuous and bounded, each $g_m(t)$ is bounded and measurable; hence, each $g_m(t)$ is in $L_2\,[a,b]$, and $\|g_m(t)\| \leq B$ for all m and $t \in [a,b]$.

We show that each subsequence of $\{g_m(t)\}$ has a further subsequence which converges to $f(t,x_o(t)) \equiv g_o(t)$ in $L_2\,[a,b]$. Let $\{g_k(t)\}$ be such an arbitrary subsequence. The corresponding sequence $\{x_k(t)\}$ has a subsequence $\{x_j(t)\}$ which converges pointwise a.e.. But then

$$\|g_j(t) - g_o(t)\|^2 = \|f(t,x_j(t)) - f(t,x_o(t))\|^2$$

converges to 0 pointwise a.e. by the continuity of f. Moreover, by the bounded convergence theorem $\int_a^b \|g_j(t) - g_o(t)\|^2 dt \to 0.$

Lemma VI.13. If rank $[C,D] = n$, then L_1 is Fredholm of index 0.

Proof. Identical to the proof of Lemma VI.6 since $L_1 = L_0$.

Remark. It is easy to verify that projections P_1, Q_1 may be defined by (VI.8) and (VI.9) and K_{P_1} may be expressed in integral form as in Lemma VI.9 with the same Green's function.

Lemma VI.14. If f is bounded and continuous, N_1 is L_1 - compact on $\bar{\Omega}$ for any open bounded $\Omega \subset X_1$.

Proof. Since f is bounded, N_1 is continuous and

$$(I - Q_1)N_1(\bar{\Omega})$$

is bounded in $L_2[a,b]$.

Let $z \in (I - Q)N_1(\bar{\Omega})$ and $x = K_{P_1} z$. Then

$$x(t) = \int_a^b h(s,t)\, z(s)ds \leq \left(\int_a^b (h(s,t))^2\, ds\right)^{\frac{1}{2}} \|z\|_2.$$

Since $h(s,t)$ is bounded, $K_{P_1}(I - Q)N_1(\bar{\Omega})$ is bounded in the uniform norm. Moreover,

$$x'(t) = \int_a^b \frac{\partial h(s,t)}{\partial t} z(s)ds.$$

Thus

$$\|x'(t)\| \leq \left(\int_a^b \frac{\partial h(s,t)^2}{\partial t} ds\right)^{\frac{1}{2}} \|z\|_2 .$$

It follows that there exists B such that

$$\|x\| + \|x'\| \leq B$$

for $x \in K_{P_1}(I - Q_1)N_1(\bar{\Omega})$. By Ascoli's theorem,

$$\overline{K_{P_1}(I - Q_1)N_1(\bar{\Omega})}$$

is compact in $L_2[a,b]$.

Let $\{\varphi_i\}$ be a complete orthonormal sequence in $L_2[a,b]$. Define

$$X_m = \text{span} < \varphi_1, \varphi_2, \ldots, \varphi_m >$$

and let T_m denote orthogonal projection of $L_2[a,b]$ onto X_m. Let $S_m{}^*$ denote the set of solutions to

$$x_m = T_m[Px_m + JQN x_m + K(I - Q) Nx_m] = M_m x_m.$$

Let

$$\rho_2(x,S) = \inf_{y \in S} \|x - y\|_2.$$

Theorem VI.9. Let f be bounded and continuous. Suppose

a) $L_1x \neq N_1x$, $x \in \partial\Omega$

b) $d[(L_1,N_1),\Omega] \neq 0$.

Then $S_m^* \neq \phi$ for m sufficiently large $S \neq \phi$ and

$$\sup_{x_m \in S_m^*} \rho_2(x_m,S) \to 0.$$

Proof. Immediate from Theorem VI.7.

In the special case where L_1 is invertible we have :

Theorem VI.10. If f is bounded and continuous and L_1 is invertible, then there exists $\Omega = B_R(0)$ such that $S_m^*(\Omega) \neq \phi$ for m sufficiently large, $S(\Omega) \neq \phi$ and

$$\sup_{x_m \in S_m} \rho_2(x_m,S) \to 0.$$

Proof. In this case there exists R such that

$$\|M_1x\| = \|Kp_1N_1(x)\| < R$$

for all $x \in L_2[a,b] = X_1$. Let $\Omega = B_R(0)$. Then $M_1\ (B_R(0)) \subset B_R(0)$ and $d[(L_1,N_1),\Omega] = d[I - M_1,\Omega,0] \neq 0$.

Theorem VI.11. In Theorems VI.9 and VI.10, if we define for each $x_m \in S_m^* y_m = Mx_m$, then

$$\sup_{x_m \in S_m^*} \rho(y_m,S) \to 0$$

where ρ is defined in terms of the maximum norm.

Suppose the assertion is not true ; i.e., suppose there exists a sequence $\{y_{m(j)}\}$ $(m(j) \to +\infty$ as $j \to +\infty)$ such that (VI.17)

$$\|y_{m(j)} - x\| > \varepsilon$$

for every $x \in S$ and all j. The corresponding sequence $\{x_{m(j)}\}$ is in Ω; thus, $\{x_{m(j)}\}$ is bounded in $L_2[a,b]$.

Thus, $P_1\ \{x_{m(j)}\}$ is bounded in $L_2[a,b]$. But $P_1\ \{x_{m(j)}\} \subset \text{Ker } L_1$. Since all norms are equivalent on Ker L_1, $P_1\{x_{m(j)}\}$ is bounded (and, hence, relatively compact) in $C[a,b]$. Similarly, $J_1Q_1N_1\ (\{x_{m(j)}\})$ is relatively compact in $C[a,b]$.

Moreover, since $N(\{x_{m(j)}\})$ is bounded in $L_2[a,b]$, $K_{P_1}(I - Q_1)\, N(\{x_{m(j)}\})$ is relatively compact in $C\{a,b\}$. Thus

$$M\{x_{m(j)}\} = \{y_{m(j)}\}$$

has a subsequence denoted by $y_{\ell(j)}$ which converges uniformly on $[a,b]$ to a continuous function $x_0(t)$. We show that $x_0(t) \in S$ which would contradict (VI.17).

We have

$$x_{\ell(j)} = T_{\ell(j)}\, y_{\ell(j)}$$

and

$$\begin{aligned} \|x_{\ell(j)} - x_0\|_2 &\leq \|T_{\ell(j)} (y_{\ell(j)} - x_0)\|_2 \\ &\quad + \|(I - T_{\ell(j)})\, x_0\|_2 \\ &\leq \|y_{\ell(j)} - x_0\|_2 + \|(I - T_{\ell(j)})\, x_0\|_2 . \end{aligned}$$

Thus $\|x_{\ell(j)} - x_0\|_2 \to 0$. Since M is continuous on $L_2[a,b]$,

$$M\, x_{\ell(j)} \to M\, x_0 .$$

But then

$$M\, x_0 = x_0 .$$

Remarks.

1) As mentioned at the outset of this section the assumption of boundedness may be weakened. In particular all the results of this section carry through under the assumption of quasi-boundedness on f.

2) In the special case of Theorem VI.10, we are left the finite dimensional equation

$$x_m = M_m\, x_m = T_m\, K N_1(x_m)$$

to solve for approximate solutions S_m^*.
Note that

$$M_m\, B_R^m(0) \subset B_R^m(0)$$

where $B_R^*(0) = B_R(0) \cap Xm$. Thus the Brouwer fixed point theorem applies. There exist well-developed algorithms for finding approximate solutions to such equations.

3) The procedure of this section has been used to approximate the solutions given by Theorem V.8.

6. We conclude this chapter by illustrating the approximation theory we have developed in the case of periodic solutions. We consider the problem treated in Theorem V.30 ; namely,

$$x' = f(t,x)$$
$$x(o) = x(T).$$

where $f : [0,T] \times R^n \to R^n$ is continuous and $f(t+T,x) = f(t,x)$. In the notation of section 3 of this chapter $A(t) \equiv 0$, $C = I$, and $D = -I$. Clearly rank $[I,-I] = n$.

We have

$$\text{Ker } L_o = \{x : x \equiv c \in R^n\}.$$

Consider the sequence

$$\{\varphi_i\}_{i=n+1}^{\infty} = \{e_1 \cos\alpha t,\ e_2 \cos\alpha t, \ldots,\ e_n \cos\alpha t,\ e_1 \sin\alpha t, \ldots,\ e_n \sin\alpha t,\ e_1 \cos 2\alpha t, \ldots,$$
$$e_n \cos 2\alpha t, \ldots\}$$

where $= \frac{2\pi}{T}$ and e_j is the j th column of the identity matrix I_n of order n.

We claim that $\{\varphi_i\}_{i=n+1}^{\infty}$ is a linearly independant sequence in $(I - P_o)$ dom L_o whose span is dense in $(I - P_o)$ dom L_o as a subspace of $H[0,T]$. If $f \in (I - P_o)$ dom L_o, $P_o(f) = \frac{I_n}{T} \int_0^T I_n f(t)dt = 0$. It is well known that

$$f(t) = \sum_{i=n+1}^{\infty} (\varphi_i, f)\varphi_i(t) + P_o(f)$$

where equality is in the L_2 - sense. Moreover, since $f'(t) \in L_2[0,T]$ and

$$\frac{I_n}{T} \int_0^T I_n f'(t)dt = 0,$$

$$f'(t) = \sum_{i=n+1}^{\infty} (\varphi_i, f)\varphi_i'(t)$$

where the convergence is again in the L_2 -sense. If

$$\sigma_m(t) = \sum_{i=n+1}^{m} (\varphi_i, f)\varphi_i,$$

then $$\|f - \sigma_m\|_2 + \|f' - \sigma_m'\|_2 \to 0.$$

Thus the span of $\{\varphi_i\}_{i=n+1}^{\infty}$ is dense in $(I - P_o)$ dom L_o. The independence is standard. We define

$$X_m = \langle \varphi_{n+1}, \varphi_{n+2}, \ldots, \varphi_m \rangle \oplus R^n$$
$$Z_m \quad \langle L\varphi_{n+1}, L\varphi_{n+2}, \ldots, L\varphi_m \rangle \oplus \text{Ker } L_o^*.$$

In this case $\text{Ker } L_o^* = \{z : z \equiv C \in R^n\}$ which has the basis $e_1, e_2, \ldots, e_n$.

Moreover,

$$\{L\,\varphi_i\} = \{\varphi_i'\}.$$

Orthonormalizing

$$\{e_1, e_2, \ldots, e_n, \varphi_{n+1}', \varphi_{n+2}', \ldots\}$$

we obtain

$$\{\frac{e_1}{\sqrt{T}}, \frac{e_2}{\sqrt{T}}, \ldots, \frac{e_n}{\sqrt{T}}, \frac{e_1 \sin\alpha t}{\sqrt{T/2}}, \ldots, \frac{e_n \sin\alpha t}{\sqrt{T/2}}, \frac{e_1 \cos\alpha t}{\sqrt{T/2}}, \ldots\}.$$

We then define R_m to be orthogonal projection onto Z_m.

For convenience of writing we consider only $m = (2j + 1)n$. The finite dimensional equations are then

$$R_m L x_m = L x_m = R_m N x_m.$$

We have

$$x_{(2j+1)n} \equiv \tilde{x}_j = a_o + \sum_{i=1}^{j} [a_i \cos i\alpha t + b_i \sin i\alpha t]$$

where $a_c, a_i, b_i \in R^n$. Thus

$$L\,\tilde{x}_j = \sum_{i=1}^{j} [i\alpha b_i \cos i\alpha t - i\alpha a_i \sin i\alpha t]$$

$$N\,\tilde{x}_j = f(t, \tilde{x}_j(t))$$

$$R_{(2j+1)n} N\,\tilde{x}_j = \frac{1}{T}\int_0^T f(t, \tilde{x}_j(t))dt + \sum_{i=1}^{j}\left[\left(\sqrt{\frac{2}{T}}\int_0^T f(s, \tilde{x}_j(s)) \sin i\alpha s\, ds\right)\frac{\sin i\alpha t}{\sqrt{T/2}} + \left(\sqrt{\frac{2}{T}}\int_0^T f(s, \tilde{x}_j(s)) \cos i\alpha s\, ds\right)\frac{\cos i\alpha t}{\sqrt{T/2}}\right]$$

Thus the approximation equations are equivalent to

$$\begin{cases} -\,i\alpha a_i = \dfrac{2}{T}\displaystyle\int_0^T f(s, \tilde{x}_j(s)) \sin i\alpha s\, ds, \quad i = 1, 2, \ldots j \\ i\alpha b_i = \dfrac{2}{T}\displaystyle\int_0^T f(s, \tilde{x}_j(s)) \cos i\alpha s\, ds, \quad i = 1, 2, \ldots j \\ 0 = \dfrac{1}{T}\displaystyle\int_0^T f(t, \tilde{x}_j(t))\, dt. \end{cases} \tag{IV.18}$$

Theorem VI.12. Suppose

a) G is a bound set relative to

$$x' = f(t,x)$$

on $[0,T]$ and $0 \in G$.

b) $\int_0^T f(s,C)\,ds \neq 0$, for $C \in \partial G$

c) $\int_0^T d[\int_0^T f(s,C)\,ds, G, 0] \neq 0$.

Let $S_i = \{x : x'(t) = f(t,x(t)),\ x(0) = x(T),\ x(t) \in \overline{G} \text{ for } t \in [0,T]\}$

$S_j = \{\tilde{x}_j : \tilde{x}_j \text{ satisfies (VI.18)},\ \tilde{x}_j(t) \in \overline{G} \text{ for } t \in [0,T]\}$

Then $S \neq \phi$, $S_j \neq \phi$ for j sufficiently large and

$$\sup_{\tilde{x}_j \in S_j} \rho(\tilde{x}_j, S) \to 0.$$

Proof. Follows immediately from Theorems V.30 and VI.8

6. Bibliographical notes concerning Chapter VI

The projection methods of Galerkin type we have employed in this chapter have an extensive literature. For an extensive bibliography and historical discussion see Krasnosel'skii, Vainikko, Zabreiko and Rutitskii ("Approximate Solution of Operator Equations", Walters-Noordhoff, 1972). In particular, Theorem VI.1 which provided the degree theoretic basis for the entire chapter can be found in that reference. Lemma VI.3 is due to Sweet ("Projection Operators for the Alternative Problem", University of Maryland, Technical Report TR 72-8 (1972)). Strasberg ("La Recherche de Solutions Périodique d'Equations Différentielles Non Linéaires", Thesis, Univ. Libre de Bruxelles, 1975) has given a theorem relating Galerkin approximation to coincidence degree which is essentially Theorem VI.4 but with a slightly different form of the conclusion.

Theorem VI.7 can be found in the book of Krasnosel'skii, et. al. cited above.

The construction of X_m and Z_m given by (VI.13) is closely related to a construction for linear equations given by Locker (Trans. Amer. Math. Soc. 154 (1971) 57-68).

The techniques in section 5 leading to Theorem VI.9 are due to Chen ("Constructive Methods for Nonlinear Boundary Value Problems", thesis, Colorado State University, 1974) although the method of proof used here is somewhat different. One of the algorithms for approximating Brouwer fixed points mentioned in the chapter is due to Jeppson (SIAM Mathematical Topics in Economic Theory and Computation (1972) 122-129). Chen has used this algorithm to compute solutions via Theorem VI.9.

VII. QUASIBOUNDED PERTURBATIONS OF FREDHOLM MAPPINGS.

1. Let $F : X \to Z$, with X, Z normed spaces.

Definition VII1. F will be said *quasibounded* if the number

$$|F| = \inf_{0 \le \rho < \infty} \{ \sup_{|x| \ge \rho} |F(x)|/|x| \}$$

is finite. In this case, $|F|$ is called the *quasinorm* of F.

Definition VII2. F will be said *asymptotic to the linear* mapping $B : X \to Z$ if

$$\lim_{|x| \to \infty} |Fx - Bx|/|x| = 0.$$

If B = 0, it will be said *asymptotic to zero.*

The following properties are easy consequences of the definitions.

Proposition VII1. *If F is quasibounded, then*

$$(\ \varepsilon > 0) (\ \rho > 0) (\ x \in X : |x| \ge \rho) : |Fx| \le (|F| + \varepsilon) |x| .$$

Proposition VII2. *If there exist constants* $\beta \ge 0$, $\gamma \ge 0$ *such that, for each* $x \in X$,

$$|Fx| \le \beta |x| + \gamma$$

then F is quasibounded and

$$|F| \le \beta .$$

Proposition VII3. *If F is asymptotic to the bounded linear mapping B, then F is quasibounded and* $|F| = |B|$

with $|B|$ *the usual norm of B. In particular, every mapping F asymptotic to zero is quasibounded with a quasinorm*

$$|F| = 0$$

and every bounded linear mapping $B : X \to Z$ *is quasibounded with a quasinorm* $|B|$. *Moreover, if a linear mapping B is quasibounded, it is bounded and its norm is equal to its quasinorm.*

2. Let $L : \text{dom } L \subset X \to Z$ be a Fredholm mapping of index zero and $N : X \to Z$ be a mapping which is L-compact on every bounded closed set of X. We shall prove the following theorem.

Theorem VII1. *Let L and N be like above and such that there exist projectors* $P : X \to X$, $Q : Z \to Z$ *with* Im P = ker L, Im L = ker Q *such that the following conditions hold :*

(a) $K_{P,Q}N : X \to X$ *is quasibounded.*

(b) *There exists* $\alpha \ge 0$ *and* $r > 0$ *such that each possible solution x of equation*

$$\Pi N x = 0$$

satisfies the relation

$$|Px| < \alpha |(I - P)x| + r .$$

(c) $(1 + \alpha)\, |K_{P,Q}N| < 1$.

(d) $J : \operatorname{Im} Q \to \ker L$ *being an isomorphism,*

$$d[JQN|\ker L, B(r) \cap \ker L, 0] \neq 0.$$

Then, $(L - N)(\operatorname{dom} L) \supset \operatorname{Im} L$, *i.e. equation*

$$Lx = Nx + w$$

has at least one solution for each $w \in \operatorname{Im} L$.

Proof. We shall apply Corollary IV.1 with

$$N^{*}x = Nx + w$$

for some $w \in \operatorname{Im} L$. We know that each equation

$$Lx = \lambda(Nx + w) \;, \; \lambda \in \,]0,1[\;, \qquad \text{(VII.1)}$$

is equivalent to

$$v = \lambda K_{P,Q}[N(u + v) + w] \qquad \text{(VII.2)}$$

$$0 = QN(u + v) \qquad \text{(VII.3)}$$

where we have written

$$u = Px, \quad v = (I - P)x .$$

Then, using (VII2) Proposition VII1 and L-compactness of N we have the following property

$$(\forall\, \varepsilon > 0)\, (\exists \gamma > 0)(\forall\, x \in X) : |K_{P,Q}Nx| \leq (|K_{P,Q}N| + \varepsilon)\, |x| + \gamma$$

and hence, for each possible solution $x = u + v$ of (VII.1),

$$(\forall\, \varepsilon > 0)\, (\exists \gamma > 0) : |v| \leq (|K_{P,Q}N| + \varepsilon)\, (|u| + |v|) + \gamma + |K_{P,Q}w| .$$

Let us fix now ε in such a way that

$$(1 + \alpha)\, (|K_{P,Q}N| + \varepsilon) < 1 ,$$

which is possible by assumption (c). If we write, for the corresponding γ,

$$\delta = \gamma + |K_{P,Q}w| ,$$

then, for each possible solution $u + v$ of (VII.1),

$$|v| \leq [1 - (|K_{P,Q}N| + \varepsilon)]^{-1}\, [(|K_{P,Q}N| + \varepsilon)|u| + \delta] \qquad \text{(VII4)}$$

Now (VII.3) and assumption (b) imply that

$$|u| < \alpha |v| + r$$

and hence for each possible solution x of (VII.1),

$$|x| \leq |u| + |v| < [1 - (1+\alpha)(|K_{P,Q}N| + \varepsilon)]^{-1} [(1+\alpha)\delta + r] = q.$$

Thus, assumption (1) of Corollary IV.1 is satisfied for $\Omega = B(q)$. Clearly, by condition (b) with $x \in \ker L$ and condition (d), assumptions of Corollary IV.1 hold and the proof is complete.

Corollary VII.1. *If* $N : X \to X$ *is quasibounded and* $|N| < 1$, *then* $I - N$ *is onto.*

Proof. Take $X = Z$, $L = I$, $\alpha = 0$ and any $r > 0$ in Theorem VII.1.

3. Let now X be a (possibly proper) subspace of $B(S,R^n)$, the space of bounded mappings from some set S into R^n, with a norm satisfying

$$|x| \geq \sup_{s \in S} |x(s)|$$

We then have the following

Theorem VII.2. *Let X be like above, L a Fredholm mapping of index zero and* $N : X \to Z$ *a mapping L-compact in each closed bounded set of X and such that* $K_{P,Q}N$ *is quasibounded for some couple of projectors* P, Q. *Suppose that the following conditions hold*

(1) *There exists* $\lambda > 0$ *such that for each* $u \in \ker L$ *and each* $s \in S$,

$$|u| \leq \lambda |u(s)|.$$

(2) *There exists* $r_1 > 0$ *such that*

$$\Pi N x \neq 0$$

for each $x \in \operatorname{dom} L$ *for which* $|x(s)| \geq r_1$ *for every* $s \in S$.

(3) $|K_{P,Q}N| < (1 + \lambda)^{-1}$

(4) $d[J Q N|\ker L, B(r_1), 0] \neq 0$ *where* $J : \operatorname{Im} Q \to \ker L$ *is an isomorphism,*

Then $(L - N)(\operatorname{dom} L) \supset \operatorname{Im} L$.

Proof. We show that the conditions of Theorem VII.1 are satisfied. If x is any solution of

$$\Pi N x = 0 ,$$

then, by assumption (2) there exists some $\sigma \in S$ for which

$$|x(\sigma)| < r_1 .$$

Hence, using assumption (1) and the condition an the norm of X,

$$|Px| \leqslant \lambda|(Px)(\sigma)| \leqslant \lambda\left[|x(\sigma)| + \left|\left[(I-P)x\right](\sigma)\right|\right] \leqslant \lambda r_1 + \lambda|(I-P)x|$$

which shows that condition (b) of Theorem VII2 holds with

$$\alpha = \lambda \ , \ r = \lambda r_1 \ .$$

Hence, by (3) condition (c) of Theorem VII1 is satisfied and the proof is complete.

Corollary VII2. *Let* X *be like in Theorem* VII2 *with moreover*

$$|x| = \sup_{s \in S} |x(s)|$$

when $x : S \to R^n$ *is a constant mapping. Let* L: dom L ⊂ X → Z *be such that* Im L *is closed and of codimension* n *and*

$$\ker L = \{x \in X: x \text{ is a constant function}\}.$$

Then, if, for some couple of projectors P, Q,

$$|K_{P,Q}N| < \frac{1}{2} \qquad \text{(VII5)}$$

and if conditions (2) *and* (4) *of Theorem* VII2 *hold, one has*

$$(L - N)(\text{dom } L) \supset \text{Im } L.$$

Proof. Obviously, condition (1) of Theorem VII2 holds with $\lambda = 1$ and the proof is complete.

With a slight supplementary assumption it is possible to give formulations of Theorem and Corollary VII2 which are independent of the choice of projectors P and Q. A preliminary lemma is needed.

Lemma VII1. *If* ΠN *is quasibounded and if* $K_{P,Q}N$ *is quasibounded for some couple of projectors* P,Q *then the same is true for any other couple.*

Proof. We have

$$K_{P',Q'}N = (I-P')K_{P,Q}N + (I-P')K_P(\Pi_Q^{-1} - \Pi_{Q'}^{-1})\Pi N \ .$$

and (vi) easily implies that both terms in the right hand side are quasibounded.

Definition VII2. A mapping N: X → Z will be said to be *L-quasi-bounded* if conditions of Lemma VII1 hold for some couple of projectors P, Q.

In this case, the *L-quasi-norm* of N will be the number

$$|N|_L = \inf |K_{P,Q}N|$$

where $|K_{P,Q}N|$ is the usual quasinorm of $K_{P,Q}N$ and the infimum is taken over all projectors P, Q such that Im P = ker L , Im L = ker Q .

It is clear that if L = I, L-quasiboundedness reduces to the usual one.

Corollary VII3. *If* N *is L-quasi-bounded, then conditions* (3)*(resp.* VII5*) of Theorem*

VII2 *(resp. Corollary* VII2) *can be replaced by*

$$|N|_L < (1+\lambda)^{-1} \quad (resp. \ |N|_L < \tfrac{1}{2}) .$$

Proof. Take $\varepsilon > 0$ such that $|N|_L + \varepsilon < (1+\lambda)^{-1}$ (resp. $< \frac{1}{2}$) and projectors P, Q such that

$$|K_{P,Q}N| \leq |N|_L + \varepsilon .$$

4. Let us come back now to an arbitrary normed space X to give a result with an assumption slightly stronger than condition (a) in Theorem 7.1 but with a weaker assumption than (b).

Theorem VII3. *Let* L *be a Fredholm mapping of index zero and* N *a mapping which is* L*-compact on each closed bounded set. Suppose that there exist projectors* P, Q *with* Im P = ker L, Im L = ker Q *and numbers* $\alpha \geq 0$, $\beta \geq 0$, $r > 0$, $s \geq 0$ *such that the following conditions hold.*

(a) $\forall \, x \in X, \ |K_{P,Q}Nx| \leq \beta |x| + s$

(b) *every possible solution* x *of equation* $\Pi Nx = 0$ *such that*

$$|(I-P)x| \leq \beta\,|x| + s$$

satisfies the inequality

$$|Px| < \alpha |(I-P)x| + r .$$

(c) $(1+\alpha)\beta < 1$.

(d) $J: \operatorname{Im} Q \to \ker L$ *being an isomorphism*

$$d\,[JQN|\ker L, \ B(r) \cap \ker L, \ 0] \neq 0 .$$

Then, $(L-N)(\operatorname{dom} L) \supset \operatorname{Im} L$.

Proof. It is similar to that of Theorem VII.1 and is left to the reader.

Remark. In all the theorems of section 2 to 4, the conclusion is always that $(L-N)(\operatorname{dom} L) \supset \operatorname{Im} L$. If we want to solve the equation

$$Lx = Fx + y ,$$

with $y \in Z$, then we can always write it in the form

$$Lx = Fx + Qy + (I-Q)y$$

and note that the mapping $x \mapsto Fx + Qy$ always has the same properties of quasiboundedness than F and that $(I-Q)y \in \operatorname{Im} L$. The only differences in the assertions is that ΠN is now $\Pi(F(.)+y)$ and JQN is $JQ(F(.)+y)$.

Also, the assumptions are independent of the sign of N and hence the same theorems hold at once for the mapping L + N.

As an application of those remarks let us prove the following

Theorem VII3'. *Let* L *be a Fredholm mapping of index zero and* N *a mapping which is* L*-compact on bounded sets. Suppose that condition* (a) *of Theorem* VII.3 *holds as well as the following assumptions :*

(b') $(\exists\, \delta > 0)\ (\forall K > 0)\ (\exists t_K > 0)\ (\forall t \geqslant t_K)$

$$(\forall w \in \ker L \cap \partial B(1))\ (\forall v \in \ker P \cap B(\delta)) :$$

$$|QN(t(w+v))| > K.$$

(c') $\beta < \delta(1+\delta)^{-1}$.

(d') $d_B[\,JQN|\ker L\ ,\ B(t_K), 0\,] \neq 0$.

Then, $(L - N)(\operatorname{dom} L) = Z$.

Proof. Let $y \in Z$. We shall apply Corollary IV1 with

$$N^* x = Nx + y.$$

Then, for each $\lambda \in\,]0,1[$ equation

$$Lx = \lambda(Nx + y) \qquad \text{(VII 5.1.)}$$

is equivalent to

$$v = \lambda K_{P,Q} N(u+v) + \lambda K_{P,Q}\, y \qquad \text{(VII 5.2.)}$$

$$0 = QN(u+v) + Qy \qquad \text{(VII 5.3.)}$$

with $u = Px$, $v = (I - P)x$. Therefore, using (VII5.2) and (a), we have, for all possible solutions u,v of (VII 5.1.) :

$$|v| \leqslant \beta|u+v| + s + |K_{P,Q}\, y|$$

$$= \beta|u+v| + \tilde{s}\ . \qquad \text{(VII 5.4.)}$$

From (VII5.3) we get

$$|QN(u+v)| = |Qy| \qquad \text{(VII 5.5.)}$$

and we shall write t^* the value of t_K in assumption (b') corresponding to $K = |Qy|$. If x is a solution to (VII5.1) such that $u = Px = 0$, then

$$|x| = |v| \leqslant \beta|x| + \tilde{s}$$

and hence, because of by (c'), $\beta < 1$,

$$|x| \leqslant (1-\beta)^{-1}\tilde{s}. \qquad \text{(VII 5.6.)}$$

If x is such that $u = Px \neq 0$, let $u = |u|\, w$, with $w = u/|u|$. Then (VII5.3) is equivalent to

$$QN(|u|\,(w + |u|^{-1} v)) + Qy = 0\ . \qquad \text{(VII 5.7.)}$$

If $|u|^{-1}|v| \leq \delta$, i.e. if

$$|v| \leq \delta |u| ,$$

then by (VII 5.5), (VII 5.7), the choice of t^* and assumption (b'),

$$|u| < t^*$$

and hence

$$|x| < t^* + \delta t^* = (1 + \delta)t^* . \qquad \text{(VII.5.8)}$$

Now if $|u|^{-1}|v| > \delta$, i.e. if

$$|v| > \delta |u| ,$$

then by (7.5.4)

$$|v| < \beta\delta^{-1} (1 + \delta) |v| + \tilde{s}$$

which implies using (c') that

$$|v| < [1 - \beta\delta^{-1} (1 + \delta)]^{-1} \tilde{s}$$

and hence

$$|x| < [(1 + \delta)^{-1} \delta - \beta]^{-1} \tilde{s} . \qquad \text{(VII.5.9)}$$

Thus, it follows from (VII 5.6.), (VII 5.8.) and (VII 5.9.) that if

$$R = \max [(1 + \delta)t^*, [(1 + \delta)^{-1} \delta - \beta]^{-1} \tilde{s}]$$

one has

$$x \notin \partial B(R)$$

for each possible solution x to (VII 5.1) and that

$$QN(u) + Qy \neq 0$$

for all $u \in \ker L \cap \partial B(R)$. The result follows then from Corollary IV. 1.

Theorem VII.3". *Let* H *be a Hilbert space with immer product* (.,.), L: dom L ⊂ H → H *a self-adjoint Fredholm mapping*, N: H → H *a mapping which is* L-*compact on bounded sets and satisfies condition* (a) *of Theorem* VI. 3. *with* P = Q *the orthogonal projector onto* ker L.

Assume moreover that

(b") $(\exists \delta > 0)\ (\forall K > 0)\ (\exists t_K > 0)\ (\forall t \geq t_K)$

$(\forall w \in \ker L \cap \partial B(1))\ (\forall v \in \operatorname{Im} L \cap B(\delta))$:

$(N(t(w + v)), w) > K$.

and that condition (c') *of Theorem* VI.3' *holds.*

Then (L - N) (dom L) = H.

Proof. L being self-adjoint one can take $P = Q : H \to H$ to be the orthogonal projector onto ker L. Proceeding like in the proof of Theorem VII.3' with similar notations we get

$$|v| \leq \beta |u + v| + \tilde{s}$$

Also, using the orthogonality of P, we get from (VII -5.3) that

$$(N(u + v),u) + (y,u) = 0$$

and then the proof continuous like in Theorem VI.3' by taking

$$K > - \inf_{w \in \ker L \cap \partial B(1)} (y,w) ,$$

and by using at the end the Poincaré-Bohl theorem to show that the required Brouwer degree is not zero.

5. We will now consider a class of mappings N which are asymptotic to zero.

Theorem VII.4. *Let* L *be a Fredholm mapping of index zero and* N *a mapping which is* L*-compact on every closed bounded set. Suppose that the following conditions hold for some couple of projectors* P,Q.

(a) *There exists* $\delta \in [0,1[$, $\mu \geq 0$ *and* $\nu \geq 0$ *such that, for each* $x \in X$,

$$|K_{P,Q}Nx| \leq \mu |x|^{\delta} + \nu .$$

(b) $(\forall$ bounded $V \subset \ker P)$ $(\exists\, t_o > 0)$ $(\forall t \geq t_o)$ $(\forall z \in V)$ $(\forall w \in \ker L \cap \partial B(1)) : QN(tw + t^{\delta}z) \neq 0$.

(c) *For some* $t \geq t_o$, $d[JQN|\ker L, B(t), 0] \neq 0$, *with* $J : \operatorname{Im} Q \to \ker L$ *an isomorphism.*

Then (L - N) (dom L) ⊃ Im L.

Proof. We shall apply Corollary IV.1 with

$$N^*x = Nx + y, \quad y \in \operatorname{Im} L.$$

Each equation

$$Lx = \lambda(Nx + y),\ \lambda \in]0,1] , \tag{VII.6}$$

is equivalent to

$$v = \lambda K_{P,Q}[N(u+v) + y] \tag{VII.7}$$

$$0 = QN(u + v) \tag{VII.8}$$

with

$$u = Px, \quad v = (I - P)x.$$

Then for each possible solution $x = u + v$ of (VII.6) we have

$$|v| \leq \mu |u + v|^{\delta} + \nu' \qquad \text{(VII.9)}$$

with

$$\nu' = \nu + |K_{P,Q}y|$$

and hence, if $|u| \neq 0$,

$$\frac{|v|}{|u|^{\delta}} \leq \mu + \frac{\mu\delta}{|u|^{1-\delta}} \left(\frac{|v|}{|u|^{\delta}}\right) + \frac{\nu'}{|u|^{\delta}} .$$

Thus, there exists $t_1 > 0$ such that if $|u| \geq t_1$,

$$\mu\delta / |u|^{1-\delta} \leq 1/2$$

and hence

$$|v| / |u|^{\delta} \leq 2(\mu + \nu' t_1^{-\delta}) .$$

If

$$V = \{z \in \operatorname{Im} L \mid |z| \leq 2(\mu + \nu' t_1^{-\delta})\} ,$$

then, by assumption (b) there exists $t_o > 0$ such that if $t \geq t_o$, $w \in \ker L \cap \partial B(1)$ and $z \in V$,

$$QN(tw + t^{\delta} z) \neq 0 .$$

Hence, if $|u| \geq \max(t_o, t_1)$, $v/|u|^{\delta} \in V$ and hence

$$QN(u+v) = QN\left(|u| \frac{u}{|u|} + |u|^{\delta} \frac{v}{|u|^{\delta}}\right) \neq 0$$

which implies by (VII.8) that $u + v$ cannot be a solution of (VII.6). Thus, each possible solution $x = u + v$ of (VII.6) is such that

$$|u| < \max(t_o, t_1) = t_2$$

and therefore by (VII.9)

$$|v| \leq \mu(t_2 + |v|)^{\delta} + \nu'$$

which implies that $|v| \leq t_3$, with t_3 the unique positive solution of equation.

$$\alpha - \mu(t_2 + \alpha)^{\delta} - \nu' = 0.$$

It is then easily checked that all conditions of Corollary IV.1 hold with $\Omega = B(t_2 + t_3)$, and the proof is complete.

6. In this section we shall suppose that $X = Z = H$, a Hilbert space, and give some corollaries of Theorem VII.5.

Definition VII.4. Let $S: \ker L \cap \partial B(1) \to R$ be a function and $\delta \in [0,1[$. S will be said to be a *weak δ-subasymptote to the mapping* F *with respect to* $\ker L$ if, for each bounded set $V \subset \operatorname{Im} L$ there exists $t_0 > 0$ such that

$$(\forall t \geq t_0)\ (\forall w \in \ker L \cap \partial B(1))\ (\forall v \in V) : (F(tw+t^\delta v),w) \geq S(w)$$

where (,) denotes the scalar product in H .

Corollary VII.4. *Let* $L: \operatorname{dom} L \subset H \to H$ *be a selfadjoint Fredholm mapping and* $F: H \to H$ *be L-compact on each closed bounded set. Suppose that condition* (a) *of Theorem* VII.4 *holds with* F *instead of* N *and that* F *has a weak* δ*-subasymptote* S *with respect to* $\ker L$. *Then, for each* $y \in H$ *such that*

$$(y,w) < S(w)$$

for each $w \in \ker L \cap \partial B(1)$, $y \in (L - F)(\operatorname{dom} L)$, *i.e. equation*

$$Lx = Fx - y \qquad \text{(VII.10)}$$

has a solution.

Proof. Let $Nx = Fx - y$. Of course N satisfies condition (a) of Theorem VII.4. Now,

$$(N(tw + t^\delta v), w) = (F(tw + t^\delta v), w) - (y, w)$$

and hence, for each bounded $V \subset \operatorname{Im} L$, there exists $t_0 > 0$ such that if $t \geq t_0$, $w \in \ker L \cap \partial B(1)$, $v \in V$,

$$(N(tw + t^\delta v),w) \geq S(w) - (y,w) \quad 0.$$

If $(w_1,\ldots,w_n)$ is an orthonormed base of ker L, then, L being self-adjoint we can take

$$Px = Qx = \sum_{i=1}^{n} (x, w_i)w_i$$

the orthogonal projector on ker L and hence, if $t \geq t_0$, $w \in \ker L \cap \partial B(1)$, $v \in V$,

$$(PN(tw + t^\delta v), w) = (N(tw + t^\delta w), w) > 0 \qquad \text{(VII.11)}$$

which implies that condition (b) of Theorem VII.4 is satisfied.
Lastly, from (VI.11) we have, if $t \geq t_0$ and $w \in \ker L \cap \partial B(1)$,

$$(PN(t\,w),w) > 0$$

and hence, using Poincaré-Bohl theorem,

$$d\,[PN|\ker L,\ B(t_o),\ 0] = d[\,I, B(t_o), 0] = 1$$

which achieves the proof.

<u>Corollary VII.5</u>. *Let* L *be a self-adjoint Fredholm mapping* $F: H \to H$ *be L-compact on each closed bounded set, satisfying condition* (a) *of Theorem* VII.4 *and suppose that*

$$(\forall k > 0)\ (\forall \text{ bounded } V \subset \operatorname{Im} L)\ (\exists t_o > 0)\ (\forall t \geq t_o)\ (\forall w \in \ker L \cap B(1))$$
$$(\forall v \in V): (F(tw + t^{\delta}v), w) \geq k .$$

Then, for each $y \in H$, *equation* (VII.10) *has at least one solution.*

<u>Proof</u>. Let $y \in H$ be given. If we take

$$k > \sup_{w \ker L \cap \partial B(1)} (y,w)$$

then the conditions of Corollary VI.4 are satisfied for the δ-subasymptote $S(w) = k$.

<u>Corollary VII.6</u>. *Let* L *be a self-adjoint Fredholm mapping and* F *be L-compact on each bounded closed set and satisfying condition* (a) *of Theorem* VII.4. *Suppose that*

$$(\forall \varepsilon > 0)\ (\forall \text{ bounded } V \subset H)\ (\exists t_o > 0)\ (\forall t \geq t_o)\ (\forall w \in \ker L \cap \partial B(1))$$
$$(\quad v \in V): |(F(tw + t^{\delta}v), w) - S^*(w)| < \varepsilon \qquad \text{(VII.12)}$$

for some $S^*: \ker L \cap \partial B(1) \to R$. *Then, if* $y \in H$ *is such that*

$$(y, w) < S^*(w) \qquad \text{(VII.13)}$$

for each $w \in \ker L \cap \partial B(1)$, *equation* (VI.10) *has at least one solution.*

<u>Proof</u>. It follows from the continuity of ΠF and from the uniform character of the limit in (VII.12) that S^* is continuous. Hence, there exists $a > 0$ such that, $\forall\ w \ker L \cap \partial B(1)$,

$$S^*(w) - (y, w) \geq \alpha .$$

Then, taking $\varepsilon = \frac{\alpha}{2}$ in (VII.12),

$$(\forall \text{ bounded } V \subset H)\ (\exists t_o > 0)\ (\forall t \geq t_o)\ (\forall w \in \ker L \cap \partial B(1))$$
$$(\forall v \in V): (F(tw + t^{\delta}v) - y, w) > S^*(w) - \frac{\alpha}{2} - (y,w) \geq \frac{\alpha}{2}$$

and the proof is achieved like in Corollary VII.4.

<u>Remarks</u>. a. *If conditions of Corollary* VII.6 *hold and if*

$$(F(x), w) < S^*(w)$$

for every $x \in H$, $w \in \ker L \cap \partial B(1)$, *then* (VII.13) *is necessary and sufficient for the solvability of* (VII.10).

Proof. If x is a solution, then

$$0 = (Lx, w) = (Fx, w) - (y, w) < S^*(w) - (y, w) ,$$

and (VII.13) is satisfied. The sufficiency follows from Corollary VII.6.

b. It is clear that Theorem VII.4 holds as well, with the same conditions for equation

$$Lx + Nx = 0$$

(the assumptions are independent of the sigh of N) and that Corollaries VII.4 to VII.6 hold with the same assertion for equation

$$Lx + Fx = y$$

because the assumptions are independent of the sign of L .

Let us come back now to normed spaces X and Z but assume that there exists an inner product space U, with inner product (,) and a normed space V, with norm $\|.\|$, such that $x \subset U$, $Z \subset U$ and $X \subset V$ topologically. Let $L : \text{dom } L \subset X \to Z$ be a linear Fredholm mapping of index zero and $F : X \to Z$ be L-compact on closed bounded sets of X. Assume also that there exists a continuous projector $Q : Z \to Z$ such that $\text{Im } L = \ker Q$ and which is orthogonal for the inner product in U.

Theorem VII.5. *Let us assume that the following conditions hold.*

i. *There exists* $\delta \in [0,1[$, $\mu,\mu',\nu,\nu' \geq 0$ *such that, for each* $x \in X$,

$$|K_{P,Q}F| \leq \mu|x|^{\delta} + \nu \ , \ |Fx| \leq \mu'|x|^{\delta} + \nu' \ .$$

ii. *There exist linear continuous mappings* $H : X \to Z$, $G : X \to X$, *a constant* $\beta \in [0,\delta]$ *with* $2\delta < 1 + \beta$, *and constants* $a > 0$, $b \geq 0$ *such that*

a. $H|\ker L$ *is an isomorphism onto* $\text{Im } Q$;

b. $G|\ker L$ *is one-to-one* ;

c. $(Hx,Nx) \geq a\|Gx\|^{1+\beta} - b$ *for each* $x \in X$.

Then, if $\beta > 0$, $L - F$ *is onto and if* $\beta = 0$ *and* $y \in Z$ *is such that*

$$\sup_{w \in \ker L \cap \text{bdry } B(1)} \|Gw\|^{-1}(Hw,Qy) < a \qquad \text{(VII.14)}$$

equation

$$Lx - Fx = y$$

has at least one solution.

Proof. We shall apply Theorem VI.4 with $Nx = Fx + y$ and show that condition (ii) imply conditions (b) and (c) of this theorem. If (b) is not satisfied, then there exist a bounded $\tilde{V} \subset \ker P$, a sequence (t_n) with $t_n > 0$, $n \in N^*$ and $t_n \to \infty$ if $n \to \infty$, a sequence (z_n) with $z_n \in \tilde{V}$, $n \in N^*$, and a sequence (w_n), with $w_n \in \ker L \cap \text{bdry } B($ such that

$$QF(t_n w_n + t_n^{\delta} z_n) + Qy = 0 \ . \qquad \text{(VII.15)}$$

By taking if necessary a subsequence, we can assume that $w_n \to w$ in X, with $w \in \ker L \cap \text{bdry } B(1)$, which implies in turn that $w_n + t_n^{\delta-1}z_n \to w$ in X when $n \to \infty$. Now, using (VII.15), we have, for each $n \in N^*$,

$$0 = t_n^{-(1+\beta)}(H(t_n w_n), QF(t_n w_n + t_n^{\delta}z_n) + Qy)$$

$$= t_n^{-(1+\beta)}(H(t_n w_n), F(t_n w_n + t_n^{\delta}z_n) + Qy) \quad \text{(by the orthogonality of Q)}$$

$$\geq t_n^{-(1+\beta)} a\|G(t_n w_n + t_n^{\delta} z_n)\|^{1+\beta} - bt_n^{-(1+\beta)} - t_n^{-(1+\beta)}(H(t_n^{\delta} z_n), F(t_n w_n + t_n^{\delta} z_n))$$

$$+ t_n^{-\beta}(Hw_n, Qy) \qquad \text{(by the use of (ii-c))}$$

$$\geq a\|G(w_n + t_n^{\delta-1} z_n)\|^{1+\beta} - bt_n^{-(1+\beta)} - k|H|\, t_n^{2\delta-1-\beta}(\mu' |w_n + t_n^{\delta-1} z_n|^{\delta} + \nu' t_n^{-\delta})$$

$+ t_n^{-\beta}(Hw_n, Qy)$ (use of the topological embedding of X and Z in U, of condition (i) and of the fact that (z_n) is a bounded sequence)

where k is a positive constant related to the above embedding and $\tilde{V}$.

Therefore, if $\beta > 0$ we obtain, if $n \to \infty$, using the fact that $X \subset V$, topologically,

$$0 \geq a\|Gw\|^{1+\beta}$$

which contradicts condition (ii-b). If $\beta = 0$, we have, similarly,

$$0 \geq a\|Gw\| + (Hw, Qy) ,$$

a contradiction with (VII.14). Thus, condition (b) of Theorem VII.4 is verified.

Now we deduce from (ii-c), if $x \in \ker L$, $J = H^*$, the adjoint of $H|\ker L$ (i.e. $(Hx, \tilde{y}) = (x, H^*\tilde{y})$ for $x \in \ker L$, $\tilde{y} \in \operatorname{Im} Q$),

$$(Hx, Fx + y) = (Hx, QF + Qy) = (x, J(QFx + Qy))$$

and, from (ii),

$$(Hx, Fx + y) \geq a\|Gx\|^{1+\beta} - b + (Hx, Qy) > 0$$

if $|x| \geq r$, with r sufficiently large. Then, using Poincaré-Bohl theorem,

$$d_B[J(QF(.) + Qy), B(r) \cap \ker L, 0] = 1$$

and the proof is complete.

7. In Theorem VII.4 condition (b) requires that

$$QN(tw + t^{\delta} z) \neq 0$$

for all $t \geq t_0$. We shall now give, under more assumptions upon N, a theorem which considerably weakens this assumption.

Theorem VI.6. *Let* L *be a Fredholm mapping of index zero and* N *a mapping which is L-compact on every closed bounded set. Suppose that the following conditions hold for some couple of projectors* P,Q .

(a) *There exists* $\nu \geq 0$ *such that, for all* $x \in X$, $|K_{P,Q}Nx| \leq \nu$.

(b) $(\forall r > 0)\ (\exists t_1 > 0)\ (\forall v \in \ker P : |v| \leq r)(\forall w \in \ker L \cap \partial B(1)) : QN(t_1 w + v) \neq 0.$

(c) $d_B[JQN|\ker L, B(t_1) \cap \ker L, 0] \neq 0.$

Then $(L - N)(\operatorname{dom} L) \supset \operatorname{Im} L.$

Proof. We shall apply Corollary IV.1 with

$$N^{*}x = Nx + y$$

for some $y \in \operatorname{Im} L$. We know that each equation

$$Lx = \lambda(Nx + y) \qquad \text{(VII.16)}$$

is equivalent to

$$v = \lambda K_{P,Q}(N(u + v) + y) \qquad \text{(VII.17)}$$

$$QN(u + v) = 0 \qquad \text{(VII.18)}$$

where we have written

$$u = Px \ , \quad v = (I - P)\,x \ .$$

From (VII.17) and assumption (a) we get, for each possible solution $x = u + v$ to (VII.16) :

$$|v| \leqslant \nu + |K_{P,Q}y| \ .$$

Now, if $|u| = 0$, surely, for each $t > 0$,

$$|u| < t \ ,$$

and if $|u| \neq 0$, (VII.18) can be written

$$QN(|u|\frac{u}{|u|} + v) = 0 \ .$$

Therefore it follows from (b) with $r = \nu + |K_{P,Q}y|$ that

$$|u| \neq t_1$$

for each possible solution $x = u + v$ to (VII.16). Therefore if we introduce the open bounded set in X

$$\Omega = \{x \in X : |Px| < t_1,\ |(I - P)x)| < \nu + |K_{P,Q}y|\}$$

we have, for each possible solution x to (VII.16), $x \notin \partial\Omega$. Now, if $a \in \ker L$ and

$$QN\,a = 0$$

then by assumption (b) and t_1 like above, $|a| \neq t_1$, i.e. $QN\,a \neq 0$ for $a \in \partial\Omega \cap \ker L$. Using assumption (c) the result now follows from Corollary IV.1.

Let now H be a Hilbert space with inner product (,) and norm $|.|_H$ and X,Z be Banach spaces with respective norms $|.|_X$ and $|.|_Z$ such that

$$X \subset Z \subset H.$$

We have the following consequence of Theorem VII.5.

Corollary VII. 7. *Let* $L : \text{dom } L \subset X \to Z$ *be linear, closed and such that* dom L *is dense in* X, Im L *closed in* Z, ker L *finite-dimensional and*

$$Z = \ker L \oplus \operatorname{Im} L.$$

Let $N : X \to Z$ *be compact and such that*

$$\sup_{x \in X} |N(x)|_Z < \infty .$$

Suppose that

$$(\forall r > 0)\ (\exists t_1 > 0)\ (\forall v \in X : |v|_X \leq r)\ (\forall w \in \ker L : |w|_H = 1) :$$

$$(N(t_1 w + v), w) > 0 \quad .$$

Then for each $h \in Z$ *such that*

$$(h, w) = 0$$

for all $w \in \ker L$, *the equation*

$$Lx - Nx = h$$

is solvable.

Proof. It clearly follows from the assumptions that L is Fredholm of index zero and that N is L-compact on bounded set of X and satisfies condition (a) of Theorem VI.6. Now we can take P = Q to be the orthogonal projector onto ker L with respect to the inner product in H and choose as an equivalent norm in X

$$\|x\| = |Px|_H + |(I - P)x|_X .$$

As

$$(N(tw + v), w) = (PN(tw + v, w)$$

for all $t \in R$, $w \in \ker L$ and $v \in X$

it follows from the assumptions that condition (b) of Theorem VI.6 holds. Also, for t_1 corresponding to $r = \nu + |K_{P,Q}\, h|$, one has

$$(Pn(t_1 w), w) > 0$$

and hence, using the Poincaré-Bohl theorem,

$$d_B[PN|\ker L, B(t_1) \cap \ker L, 0] = 1$$

which completes the proof.

8. Bibliographical notes concerning chapter VII

The concept of quasibounded mapping has been introduced by Granas (*Bull. Acad. Polon. Sci.* 9(1957) 867-871; *Rozpravy Mat.* 20(1962) 1-93) and the one of mapping asymptotic to a linear one by Krasnosel'skii (*Uspehi Math. Nauk* 9(1954) 57-114). Theorem VII.1 is essentially given in Mawhin (*J. Math. Anal. Appl.* 45(1974) 455-467) and Corollary VII.1 has been first proved by Granas(*op. citae)* and can be traced, in a less general form, to Dubrovskii (*Ucen. Zap. Moskow, Gos. Univ.* 30 (1939) 49-60). Theorem VII.2 is a slight extension of Corollary VII.2 which is proved in Mawhin (*op. cit.*). Lemma VII.1, Corollary VII.3 and Theorem VII.3 are given here for the first time. A special case of Theorem VII.3, with $\beta = 0$, is given in Cronin (*J. Differential Equations* 14 (1973) 581-596). Theorem VII.4 which is given in Mawhin (*Proc. Symp. Dynam. Systems*, Acad. Press, 1976, to appear) is a generalization of a result due to Fučik, Kučera and Nečas (*J. Differential Equations* 17 (1975) 375-394) which considered the case of $X = Z = H$, a Hilbert space, L self-adjoint and with stronger conditions (b) and (c). The concept of weak δ-subasymptote to a mapping with respect to a subspace is essentially due to Fučik, Kučera and Nečas (*op. cit.*) as well as Corollaries VII.5 and VII.6 which extend results of Necas (*Comment. Math. Univ. Carolinae* 14 (1973) 63-72). For another result in the same spirit, but by a quite different approach, see de Figueiredo ("On the range of nonlinear operators with linear asymptotes which are not invertible", *Comment. Math. Univ. Carolinae* 15 (1974) 415-428). Theorems VII.3' and VII.3" generalize results of Fučik (*Funkcial Ekvacioj* 17 (1974) 73-83). Theorem VII.5 is due to Fabry and Franchetti (*J. Differential Equations*, to appear) who deduced it directly from Corollary IV.1. The proof given here is in Mawhin (*Proc. Symp. Dynam. Syst.*, Acad. Press, 1975, to appear). Theorems VI.6 and Corollary VII.7 are new and generalize results of Fučik (*Comm. Math Univ. Carolinae* 15 (1974) 259-271). For related results see also Fučik (*Czechoslovak Math. J.* 24(99) (1974) 467-495, Osborn and Sather (*J. Differential* Equations 17 (1975 12-31).

VIII. BOUNDARY-VALUE PROBLEMS FOR SOME SEMILINEAR ELLIPTIC PARTIAL DIFFERENTIAL EQUATIONS.

1. Let $\Omega \subset R^n$ be a bounded domain, $a_{ij} : \Omega \to R$ (i, j = 1, ..., n) measurable and bounded functions and suppose that there exists $0 < \mu \leq M$ such that, for each $x \in \Omega$ and each $\xi \in R^n$,

$$\mu|\xi|^2 \leq \sum_{i,j=1}^{n} a_{ij}(x)\xi_i\xi_j \leq M|\xi|^2$$

with $|\xi|$ the Euclidian norm of ξ and let $f : \overline{\Omega} \times R \to R$ be a continuous function. Let us denote by $H^1 = H^{1,2}(\Omega)$ the completion of the space $\mathcal{C}^1(\Omega)$ of real $\mathcal{C}^1$-functions in Ω for the Sobolev norm

$$\|u\|_{1,2} = \|u\|_{L^2(\Omega)} + \sum_{i=1}^{n} \|D_i u\|_{L^2(\Omega)}$$

where $D_i u = \dfrac{\partial u}{\partial x_i}$ and

$$\|v\|_{L^2(\Omega)} = \left[\int_\Omega v^2(x)\, dx\right]^{\frac{1}{2}}$$

We are interested in the existence of $u \in H^1$ such that, for each $v \in H^1$, one has

$$a(u,v) = \int_\Omega \left[\sum_{i,j=1}^{n} a_{ij}(x) D_i u(x) D_j v(x)\, dx\right] = \int_\Omega f(x,u(x))v(x)dx \qquad \text{(VIII.1)}$$

If the boundary $\partial\Omega$ of Ω and the functions a_{ij}, f are sufficiently regular, one can show that (VIII.1) is equivalent to the semilinear Neumann problem.

$$-\sum_{i=1}^{n} D_i\left(\sum_{j=1}^{n} a_{ij}(x)D_j u\right) = f(x,u) \ , \ x \in \Omega$$

$$\sum_{i,j=1}^{n} a_{ij}(x)X_j(x)D_i u = 0 \ , \qquad x \in \partial\Omega$$

where the $X_j(x)$ are the components of the unit exterior normal to $\partial\Omega$ at x. We shall first write (VII.1) as a problem of the form (II.1).

2. Let

$$\text{dom}\, \tilde{L} = \{u \in H^1 : v \mapsto a(u,v) \text{ is continuous in } H^1 \text{ with the } L^2 \text{ norm}\}$$

Using the fact that H^1 is dense in $L^2(\Omega)$ and the theorem of structure of functionals in a Hilbert space, we have, for each $u \in \text{dom}\,\tilde{L}$ and each $v \in H^1$,

$$a(u,v) = (\tilde{L}u, v)$$

where (,) is the scalar product in $L^2(\Omega)$ and where $\tilde{L} : \text{dom}\,\tilde{L} \subset L^2(\Omega) \to L^2(\Omega)$

is linear (but not continuous). Hence, if $h \in L^2(\Omega)$, each equation in H^1 of the type

$$a(u,v) = (h,v) \ , \forall \ v \in H^1, \qquad \text{(VIII.2)}$$

is equivalent to

$$(\tilde{L}u,v) = (h,v) \ , \forall \ v \in H^1$$

and hence, H^1 being dense in $L^2(\Omega)$, to

$$\tilde{L}u = h \ . \qquad \text{(VIII.3)}$$

Now, it is a classical result of the L^2- theory of linear elliptic boundary value problems that under assumptions listed above, (VIII.2) (or VIII.3) is solvable if and only if h satisfies the relation

$$\int_\Omega h = 0 \ ,$$

in which case two solutions of (VIII.3) will always differ by a constant. In other words, if we define the projector $\tilde{P} : L^2(\Omega) \to L^2(\Omega)$ by

$$\tilde{P}u = (\text{meas } \Omega)^{-1} \int_\Omega u \ ,$$

then

$$\ker \tilde{L} = \operatorname{Im} \tilde{P} \ , \ \operatorname{Im} \tilde{L} = \ker \tilde{P} \ .$$

Moreover, we also know by the linear theory that there exists $k_1 > 0$ such that, for each $v \in L^2(\Omega) \cap \ker \tilde{P}$,

$$\|K_P v\|_{H^1} \leq k_1 \|v\|_{L^2(\Omega)} \qquad \text{(VIII.4)}$$

On the other side, it follows from regularization theory for (VIII.2) that if Ω satisfies some regularity assumptions that we shall not explicit here and if $h \in L^p(\Omega)$ with $p > n$, then each solution $u \in H^1$ of (VIII.2) or (VIII.3) is Hölder-continuous with some coefficient $\alpha \in \,]0,1[$ and there exists $k_2 > 0$ independent of u such that

$$\|u\|_{C^{0,\alpha}(\Omega)} \leq k_2(\|u\|_{L^2(\Omega)} + \|h\|_{L^P(\Omega)})$$

with

$$\|u\|_{C^{0,\alpha}(\Omega)} = \sup_{x\in\Omega} |u(x)| + \sup_{\substack{x,y\in\Omega \\ x\neq y}} \frac{|u(x)-u(y)|}{\|x-y\|^\alpha} \ .$$

From this result and from the compactness of the canonical injection $i : C^{0,\alpha}(\Omega) \to C^0(\bar{\Omega})$, with $C^0(\bar{\Omega})$ the space of real continuous functions on $\bar{\Omega}$ we see that if L is the restriction of $\tilde{L}$ to $L^{-1}(C^0(\bar{\Omega}))$ and P the restriction of $\tilde{P}$ to $C^0(\bar{\Omega})$, then

$$\ker L = \operatorname{Im} P \ , \ \operatorname{Im} L = \ker P$$

and there exists $k_3 > 0$ such that, for each $v \in \ker P$,

$$\|K_P v\|_{C^{0,\alpha}(\Omega)} \leqslant k_3 \|v\|_{C^0(\overline{\Omega})} ,$$

which implies the compactness of K_P.

3. If now we define $N : C^0(\overline{\Omega}) \to C^0(\overline{\Omega})$ by

$$(Nu)(x) = f(x\ ,\ u(x)),\ x \in \overline{\Omega}\ ,$$

it is easily checked that N is continuous and takes bounded sets into bounded sets. Hence, L is a Fredholm mapping of index zero and N is L-compact on each bounded set, and the solutions of (VIII.1) which belong to $C^0(\overline{\Omega})$ are the solutions of the operator equation

$$Lu = Nu$$

in $C^0(\overline{\Omega})$.

Now if there exists $\beta \geqslant 0$, $s \geqslant 0$ such that, for each $x \in \overline{\Omega}$ and each $u \in R$,

$$|f(x,u)| \leqslant \beta |u| + s\ , \qquad \text{(VIII.5)}$$

we get at once that

$$|Nu|_{C^0(\overline{\Omega})} \leqslant \beta |u|_{C^0(\overline{\Omega})} + s$$

and hence

$$|K_{P,P} Nu|_{C^0(\overline{\Omega})} \leqslant 2k_3 (\beta |u|_{C^0(\overline{\Omega})} + s)$$

which implies that the quasinorm of $K_{P,P}N$ is smaller or equal to $2k_3\beta$.

. Theorem VIII.1. *Let* $\Omega \subset R^n$ *and* $a_{ij} : \Omega \to R$ *be like above and let* $f : \Omega \times R \to R$ *be a continuous function verifying* (VIII.5). *Suppose that the following conditions are satisfied.*

(a) $\beta < \frac{1}{4} k_3$ (VIII.6)

(b) *There exists* $R > 0$ *such that,* $\forall\, u \in C^0(\overline{\Omega})$ *such that,* $\forall\, x \in \overline{\Omega}$, $|u(x)| \geqslant R$, *we have*

$$\int_\Omega f(x,u(x))dx \neq 0$$

(c) $\left[\int_\Omega f(x,-R)dx\right]\left[\int_\Omega f(x,R)dx < 0\right].$

Then the problem (VIII.1) *has at least one solution belonging to* $C^{o}(\bar{\Omega})$.

<u>Proof</u>. The preceding discussion shows that the couple (L,N) defined in sections 2 and 3 satisfies the basic conditions of Corollary VII.2. As (c) implies that condition (4) of this Corollary is also satisfied, the proof is complete.

<u>Remarks</u>. 1) Condition (a) above will be satisfied in the case where

$$\frac{|f(x,u)|}{|u|} = 0 \quad \text{if} \quad |u| \to \infty$$

uniformly in $x \in \bar{\Omega}$.

2) Condition (b) above will be satisfied if $f(x,u) \neq 0$ for all $x \in \bar{\Omega}$ and $|u| \geq R$.

<u>Corollary VIII.1</u>. *Suppose that* f *satisfies* (VIII.5), (VIII.6) *and is of the form*

$$f(x,u) = h(x) - g(u)$$

where $h : \bar{\Omega} \to R$ *and* $g : R \to R$ *are continuous, that*

$$g_+ = \lim_{u\to+\infty} g(u) \quad , \quad g_- = \lim_{n\to-\infty} g(u)$$

exist (possibly infinite), and that, for each $u \in R$ *either*

$$g_- < g(u) < g_+ \tag{VIII.7}$$

or

$$g_+ < g(u) < g_-$$

Then a necessary and sufficient condition for the existence of one solution of (VIII.1) *belonging to* $C^{o}(\bar{\Omega})$ *is that either*

$$g_- < h_o < g_+$$

or

$$g_+ < h_o < g_-$$

with

$$h_o = (\text{meas } \Omega)^{-1} \int_\Omega h.$$

<u>Proof</u>. Let us consider the case where (VIII.7) is satisfied. If u is a solution (VIII.1) then, by taking $v(x) \equiv 1$ in (VIII.1) we get

$$0 = h_o - (\text{meas } \Omega)^{-1} \int_\Omega g(u(x))dx$$

and hence

$$g_- < h_o < g_+ ,$$

and the necessity is proved. For the sufficiency, consider, say, the case where g_+

is finite and $g_- = -\infty$; there exists $R_1 > 0$ such that, if $|u| \geq R_1$,

$$g_+ - g(u) \leq \frac{1}{2}(g_+ - h_o)$$

and hence, if $|u(x)| \geq R_1$ for each $x \in \bar{\Omega}$,

$$h_o - (\text{meas } \Omega)^{-1} \int_\Omega g(u(x))dx \leq h_o - g_+ + \frac{1}{2}(g_+ - h_o) =$$

$$= \frac{1}{2}(h_o - g_+) < 0 .$$

Also, there exists $R_2 > 0$ such that, if $|u| \geq R_2$,

$$h_o - g(u) > 0$$

and hence, if $|u(x)| \geq R_2$ for each $x \in \bar{\Omega}$,

$$h_o - (\text{meas } \Omega)^{-1} \int_\Omega g(u(x))dx > 0 .$$

Thus, conditions (b) and (c) of Theorem VIII.1 hold with $R = \max(R_1, R_2)$ and the proof is complete.

5. Let Ω be a bounded domain in R^n and $a_{\alpha\beta}$, $0 \leq |\alpha|, |\beta| \leq m$ be real valued $C^\infty(\Omega)$-functions with $a_{\alpha\beta}$, for $|\alpha| = |\beta| = m$, uniformly continuous. As usual, $\alpha = (\alpha_1, \ldots, \alpha_n)$, $\alpha_i \in N$, and $|\alpha| = \sum_{i=1}^n \alpha_i$. Let us suppose that $a_{\alpha\beta} = a_{\beta\alpha}$ and that there exist constant $c > 0$ such that

$$\sum_{\substack{|\alpha|=m \\ |\beta|=m}} a_{\alpha\beta}(x)\, \xi^\alpha \xi^\beta \geq c|\xi|^{2m}$$

for all $\xi \in R^n$ and all $x \in \Omega$. Let $H_o^m(\Omega)$ be the completion of the space $C_o^\infty(\Omega)$ for the norm

$$\|\phi\|_m = \left[\sum_{|\alpha| \leq m} \int_\Omega |D^\alpha \phi|^2 \right]^{\frac{1}{2}}$$

and let us define the bilinear form

$$a(u,v) = \sum_{\substack{|\alpha| \leq m \\ |\beta| \leq m}} \int_\Omega a_{\alpha\beta}(x) D^\alpha u(x) D^\beta v(x)\, dx .$$

Let

$$f : \Omega \times R^n \to R ,$$

be a function satisfying Caratheodory conditions, i.e. (i) for each fixed $u \in R^n$, the function $x \to f(x,u)$ is measurable in Ω, (ii) for fixed $x \in \Omega$ (a.e.) the function $u \to f(x,u)$ is continuous in R^n. Let us also suppose that there exist constants $c > 0$, $0 \leq \beta < 1$ and a function $d \in L^2(\Omega)$ such that for $x \in \Omega$ (a.e.),

$$|f(x,s)| \leq c|s|^{\beta} + d(x).$$

This implies in particular that the mapping N defined by

$$(Nu)(x) = f(x,u(x))$$

is a continuous mapping from $L^2(\Omega)$ into itself and that

$$|Nu|_{L^2} \leq c|u|^{\beta}_{L^2} + |d|_{L^2} . \tag{VIII.8}$$

We will be interested in finding $u \in H^m_o$ such that, for each $v \in H^m_o$,

$$a(u,v) = \int_{\Omega} f(x,u(x))v(x)dx \tag{VIII.9}$$

If we denote by dom L the subspace of H^m_o

$$\{u \in H^m_o : v \mapsto a(u,v) \text{ is continuous in } H^m_o \text{ with the } L^2\text{-norm}\}$$

we obtain as in section 2 that there exists an unique linear mapping $L : \operatorname{dom} L \subset L^2(\Omega) \to L^2(\Omega)$ such that

$$a(u,v) = (Lu,v)$$

and hence (VIII.9) is equivalent to the equation in $L^2(\Omega)$

$$Lu = Nu.$$

Also it is known from L^2-theory of elliptic boundary value problems that L is a Fredholm mapping of index zero which has a compact right inverse. We will be interested here in the case where $\ker L \neq \{0\}$ and prove the following result.

6. Theorem VIII.2. *Suppose that the assumptions of section 5 hold for* $a(u,v)$ *and* $f(x,u)$. *If there exist functions* $h_+ \in L^{2/(1-\beta)}$, $h_- \in L^{2/(1-\beta)}$ *such that*

$$\lim_{s \to \pm\infty} \frac{f(x,s)}{|s|^{\beta}} = h_{\pm}(x)$$

and if for all $v \in \ker L \cap \partial B(1)$ *one has*

$$\int_{\Omega_+} h_+|v|^{1+\beta} - \int_{\Omega_-} h_-|v|^{1+\beta} > 0 \tag{VIII.10}$$

with $\Omega_{\pm} = \{x \in \Omega : v(x) \gtrless 0\}$, *then* (VIII.9) *has at least one solution.*

Proof. Let us show first that

$$(\forall \text{ bounded } V \subset \operatorname{Im} L)\ (\exists t_o > 0)\ (\forall t \geq t_o)\ (\forall v \in V)$$
$$(\forall w \in \ker L \cap \partial B(1)) : (N(tw + t^{\beta}z), w) > 0 . \tag{VIII.11}$$

If not,

$$(\exists \text{ bounded } V \subset \operatorname{Im} L)\ (\exists (t_n),\ t_n > 0,\ t_n \to \infty \text{ if } n \to \infty)(\exists (w_n),$$
$$w_n \in \ker L \cap \partial B(1))\ (\exists (v_n),\ v_n \in V) : \int_{\Omega} f(x, t_n w_n(x) + t_n^{\beta} v_n(x)) w_n(x)\,dx \leq 0$$
$$n = 1, 2, \ldots \tag{VIII.12}$$

By going if necessary to subsequences, we can suppose that $w_n \to w$ in $L_2(\Omega)$, $w_n + t_n^{\beta-1} v_n \to w$ in $L_2(\Omega)$, $w_n(x) \to w(x)$ and $w_n(x) + t_n^{\beta-1} v_n(x) \to w(x)$ a.e. in Ω. Hence, for almost each $x \in \Omega_+$ (resp. Ω_-) there exists $n_o(x) > 0$ such that, if $n \geq n_o(x)$, $w_n(x) + t_n^{\beta-1} v_n(x) > \frac{w(x)}{2}$ (resp. $< \frac{w(x)}{2}$) which implies that, a.e. in Ω_+ (resp. Ω_-),

$$t_n w_n(x) + t_n^{\beta} v_n(x) \to +\infty \quad (\text{resp. } -\infty)$$

if $n \to \infty$. Now, for each $n = 1, 2, \ldots$,

$$0 \geq \int_{\Omega} t_n^{-\beta} f(x, t_n w_n(x) + t_n^{\beta} v_n(x)) w_n(x)\,dx$$
$$= \int_{\Omega} t_n^{-\beta} f(\ldots) w(x)\,dx + \int_{\Omega} t_n^{-\beta} f(\ldots)(w_n(x) - w(x))\,dx$$
$$= \int_{\Omega_+} t_n^{-\beta} f(\ldots) w(x)\,dx + \int_{\Omega_-} t_n^{-\beta} f(\ldots) w(x)\,dx + \int_{\Omega} t_n^{-\beta} f(\ldots)(w_n(x) - w(x))\,dx$$
$$= I_+ + I_- + II. \tag{VIII.14}$$

Now,

$$\left|\int_{\Omega} t_n^{-\beta} f(\ldots)(w_n(x) - w(x))\,dx\right| \leq \left|\int_{\Omega} (t_n^{-2\beta} f^2(\ldots)\,dx)^{\frac{1}{2}}\right| |w_n - w|_{L^2}$$
$$\leq \left[c|w_n + t_n^{\beta-1} v_n|^{\beta}_{L_2} + t_n^{-\beta} |d|_{L_2}\right] |w_n - w|_{L^2} \leq M |w_n - w|_{L^2}$$

using the fact that $(w_n + t_n^{\beta-1} v_n)$ converges in L^2 to w. Thus, integral (II) goes to zero if $n \to \infty$. On the other side, the sequence $\left(\frac{f(\cdot, t_n w_n(.) + t_n^{\beta} v_n(.))}{t^{\beta}}\right)$, bounded in $L^2(\Omega)$, and hence in $L^2(\Omega_\pm)$, converges weakly in $L^2(\Omega_\pm)$ to its pointwise limit which is, using (VIII.12),

$$\lim_{n\to\infty} \frac{f(x, t_n w_n(x) + t_n^\beta v_n(x))}{t_n^\beta} =$$

$$\lim_{n\to\infty} \frac{f(x, t_n w_n(x) + t_n^\beta v_n(x))}{|t_n w_n(x) + t_n^\beta v_n(x)|^\beta} \cdot |w_n(x) + t_n^{\beta-1} v_n(x)|^\beta$$

$$= h_\pm(x)|w(x)|^P \quad , \text{ a.e. in } \Omega_\pm .$$

Hence

$$\int_{\Omega_\pm} t_n^{-\beta} f(\ldots)w(x)dx \to \pm \int_{\Omega_\pm} h_\pm(x)|w(x)|^{1+\beta}dx$$

Going to the limit in (VIII.14) we then obtain

$$0 \geq \int_{\Omega_+} h_+(x)|w(x)|^{1+\beta}dx - \int_{\Omega_-} h_-(x)|w(x)|^{1+\beta}\,dx$$

which contradicts (VIII.10). Thus, (VIII.11) holds and, using the symmetry of L we have that

$$\ker L = \operatorname{Im} P \quad , \quad \ker P = \operatorname{Im} L$$

with P the orthogonal projector onto $\ker L$, which together with (VIII.11), shows that assumption (b) of Theorem VII.4 holds, assumption (a) following from (VIII.8). On the other hand, taking $v = 0$ in VIII.11 and using the orthogonal character of P we have

$$(PN(tw), w) > 0$$

for all $t \geq t_0$ and all $w \in \ker L$ which implies, by Poincaré-Bohl theorem, that

$$d[PN|\ker L, B(t_0), 0] = 1 \ ,$$

i.e. that condition (c) of Theorem VII.4 is satisfied, which completes the proof.

<u>Corollary VIII.2</u>. *Suppose that assumptions of Theorem* VIII.2 *hold with* $\beta = 0$. *Then if a.e. in* Ω *and for each* $u \in R$,

$$h_-(x) \leq f(x,u) \leq h_+(x) \tag{VIII.15}$$

condition (VIII.10) *with* $\beta = 0$ *and non strict inequality signs is necessary for the existence of one solution for* (VIII.9).

Proof. Sufficiency has been proved. Now if u verifies (VIII.9), then taking $v \in \ker L$ and using the symmetry of a, we get

$$\int_{\Omega} f(x,u(x))v(x)dx = 0 .$$

i.e.

$$\int_{\Omega_+} f(x,u(x))|v(x)|\,dx - \int_{\Omega_-} f(x,u(x)|v(x)|\,dx = 0$$

Therefore, using VIII.15, we get

$$\int_{\Omega_+} h_+(x)|v(x)|\,dx - \int_{\Omega_-} h_-(x)|v(x)|\,dx \geqslant 0 ,$$

which achieves the proof.

7. Let $\Omega \subset \mathbb{R}^n$ be a bounded domain with smooth boundary Γ and L be a linear elliptic partial differential operator of order m with smooth coefficients acting on scalar functions and satisfying "coercive" (Lopatinsky-Shapiro) smooth boundary conditions.

$$Bu = 0 \quad \text{on} \quad \Gamma$$

expressed in terms of $m/2$ differential operators of order $< m$. Then the operator L acting on such functions is of Fredholm type and we shall suppose that its index is zero. We shall consider the boundary value problem

$$Lu = f(x,u) \text{ in } \Omega \text{ , } Bu = 0 \quad \text{on } \Gamma , \qquad \text{(VIII.16)}$$

with $f : \Omega \times R \to R$ continuous and having limits as $u \to \pm\infty$; for simplicity we shall suppose

$$\lim_{u \to \pm\infty} f(x,u) = h_{\pm}(x)$$

uniformly for x in Ω. Also we shall make the following hypothesis on unique continuation of elements in ker L :

(UC) *The only solution of*

$$Lw = 0 \text{ , } Bw = 0 \text{ on } \Gamma$$

which vanishes on a set of positive measure in Ω *is* $w = 0$.

The solutions of (VIII.16) are to be understood as functions belonging to $H^{m,p}(\Omega)$, i.e. having generalized derivatives up to the order m which belong to L^p, for every $p < \infty$. Let $w_1, \ldots, w_d$ (resp. $w_1', \ldots, w_d'$) be smooth functions spanning $\ker L \cap \ker B$ (resp. $(\operatorname{Im} L \cap \ker B)^{\perp}$). If $a = (a_1,\ldots,a_d)$ is any vector in R^d, denote

$$\sum_{i=1}^{d} a_i w_i = a.w$$

and define $\phi : R^d \to R^d$ by

$$\phi_i(a) = \int_{a.w>0} h_+(x)w_i'(x)dx + \int_{a.w<0} h_-(x)w_i'(x)dx \ , \ i = 1,\dots,d$$

and for $u \in L^1(\Omega)$, define

$$\phi_i(u,ra) = \int_{\Omega} f(x, ra.w(x) + u(x))w_i'(x)dx \ , \ i = 1,\dots,d$$

8. Lemma VIII.1. *Suppose the assumptions of section 7 are satisfied. Then*

$$\lim_{r\to\infty} | \Phi_i(u,ra) - \phi_i(a) | = 0 \ (r > 0)$$

uniformly for u *bounded in* $L^1(\Omega)$ *and* $a \in S^{d-1}$.

Proof. First step. Let $w \in L^1(\Omega)$, $w \neq 0$ a.e. Then,

$$\lim_{c\to 0} \text{meas}\{x : |w(x)| \leq c\} = 0 \ .$$

In fact, if

$$A_n = \{x : |w(x)| < \tfrac{1}{n}\} \ ,$$

A_n is a nonincreasing sequence of measurable sets. By the monotone convergence theorem applied to the characteristic functions of the A_n, we have

$$\text{meas}\ (\cap_n A_n) = \lim_{n\to\infty} \text{meas}\ A_n \ .$$

But $\cap_n A_n$ is the set where w vanishes, and hence has measure zero, so the lemma is proved.

Second step. Let the assumption (UC) above hold. Then, if

$$m(c) = \sup_{|a|=1} \text{meas}\ \{x : |a.w(x)| \leq c\}$$

we have

$$\lim_{c\to 0} m(c) = 0 \ .$$

If not there will exist sequences $c_k \to 0$ and $a_k \in S^{d-1}$ such that

$$\text{meas}\ A^k \geq \varepsilon > 0 \ \text{ with } A^k = \{x : |a^k.w(x)| \leq c_k\}.$$

By taking if necessary a subsequence, we may assume that $a_k \to a$; let $c_k' = |a_k - a|$.

Then, because of the relations

$$c_k \geq |a_k.w(x)| = |a.w(x) + (a_k-a).w(x)| \geq |a.w(x)| - |a_k-a| \, |w(x)| ,$$

we have,

$$A^k \subset \{x : |a.w| \leq c_k + c_k'|w(x)|\}$$

$$\subset \{x : |a.w| \leq c_k + c_k'\mu\} \cup \{x : |w(x)| > \mu\}$$

where $\mu > 0$. Hence,

$$\varepsilon \leq \text{meas}\,\{x : |a,w| \leq c_k + c_k'\mu\} + \frac{|w|_{L^1}}{\mu} .$$

Choosing first μ so that $|w|_{L^1}/\mu < \varepsilon/2$ and then k so that

$$\text{meas}\,\{x : |a.w| \leq c_k + c_k'\mu\} < \frac{\varepsilon}{2} ,$$

which is allowed by the first step, we obtain a contradiction and the second step is proved.

Third step. We have

$$|\Phi_i(u,ra) - \phi_i(a)| \leq \int_{a.w>0} |\left[f(x,ra.w(x)+u(x)-h_+(x)\right] w_i'(x)| \, dx$$

$$+\int_{a.w<0} |\left[f(x,\ ra.w(x) + u(x)) - h_-(x)\right] w_i'(x)| \, dx = I_1 + I_2 .$$

We shall estimate the first integral, the method is the same for the second one. By the assumptions, there exists $k > 0$ such that

$$|f(x,u)| \leq k , \ x \in \Omega, \ u \in R .$$

Let $\varepsilon > 0$; then if $\eta < \varepsilon/k$, we have, for each set A such that $\text{meas}\, A < \eta$,

$$\int_A k < \varepsilon ,$$

and, by Egorov theorem, there exists $\Omega_\varepsilon \subset \Omega$ with $\text{meas}\,(\Omega-\Omega_\varepsilon)<\eta$ such that

$$f(x,\ ra.w(x) + u(x))w_i'(x) \to h_+(x)w_i'(x)$$

uniformly in $x \in \Omega_\varepsilon$. Hence, there exist $N_\varepsilon > 0$ such that, if $|ra.w(x) + u(x)| \geq N_\varepsilon$, $x \in \Omega_\varepsilon$,

$$|[f(x,\ ra.w(x) + u(x)) - h_+(x)]w_i'(x)| \leq \varepsilon .$$

Thus,

$$I_1 \leqslant \int_{\substack{a.w>0 \\ \Omega\setminus\Omega_\varepsilon}} + \int_{\substack{a.w>0 \\ \Omega_\varepsilon \\ |ra.w(x)+u(x)| \geqslant N_\varepsilon}} + \int_{\substack{a.w>0 \\ \Omega_\varepsilon \\ |ra.w(x)+u(x)| < N_\varepsilon}}$$

$$= I_1' + I_1'' + I_1''' \leqslant 2\varepsilon + \varepsilon(\text{meas } \Omega) + I_1''' .$$

If $|u|_{L^1} \leqslant \rho$, then, if we set $\mu = \rho/\eta$, we have,

$$\rho = \int_\Omega |u| = \int_{|u(x)| \leqslant \mu} |u| + \int_{|u(x)| > \mu} |u|$$

and hence

$$\text{meas } \{x : |u(x)| > \mu\} < \rho/\mu = \eta$$

Thus

$$I_1''' \leqslant \int_{\substack{a.w>0 \\ \Omega_\varepsilon \\ |ra.w(x)| < N_\varepsilon + |u(x)|}}$$

$$\leqslant \int_{\substack{a.w>0 \\ \Omega_\varepsilon \\ |ra.w(x)| < N_\varepsilon + \mu}} + 2\varepsilon .$$

But, by the second step,

$$\text{meas } \{x : |ra.w| \leqslant N_\varepsilon + \mu\} \leqslant m\left(\frac{N_\varepsilon + \mu}{r}\right)$$

and we may chose r so large that

$$m\left(\frac{N_\varepsilon + \mu}{r}\right) < \eta ,$$

which implies that

$$\int_{\substack{a.w>0 \\ \Omega_\varepsilon \\ |ra.w(x)+u(x)| < N_\varepsilon}} \leqslant 4\varepsilon ,$$

and achieves the proof.

9. Proposition VII.1. *Under conditions of section* 7, *the mapping* $\phi : S^{d-1} \to R^{d-1}$ *is continuous.*

Proof. For each $i = 1, \ldots, d$, and each $u \in L^1$ the mapping $a \to \Phi_i(u,ra)$ is continuous and, by Lemma VIII.1, $\Phi_i(u,ra)$ converges to $\phi_i(a)$ uniformly on S^{d-1} as $r \to \infty$. This implies that ϕ is continuous.

10. Let us now formulate problem (VIII.16) as an operator equation. Take $X = C^0(\overline{\Omega})$ with the uniform norm, and, for some fixed $p > n$, dom $\tilde{L} = \{u \in H^{m,p}(\Omega),\ Bu = 0\}$. It is known that dom $\tilde{L} \subset C^0(\overline{\Omega})$, with the canonical injection compact. Let

$$\tilde{L} : \text{dom } \tilde{L} \to L^p(\Omega)\ ,\ u \to Lu$$

$$N : C^0(\overline{\Omega}) \to C^0(\overline{\Omega})\ ,\ u \to f(.,u(.))$$

$$\text{dom } L = \tilde{L}^{-1}(C^0(\overline{\Omega}))\ ,$$

$$L : \text{dom } L \to C^0(\overline{\Omega})\ ,\ u \to Lu\ .$$

It follows from the regularity theory of linear elliptic problems that (L,N) satisfies the required Fredholm and compactness assumptions and that each solution in dom L of

$$Lu = Nu \tag{VIII.17}$$

is a solution of (VIII.16)

11. Theorem VIII.3. *Assume that conditions of section* 7 *hold. Then, if*

$$\phi(a) \neq 0\ ,\ a \in S^{d-1} \tag{VIII.18}$$

and if

$$d[\tilde{\phi}, B(1), 0] \neq 0$$

where $\tilde{\phi}$ *is any continuous extension of* $\phi | S^{d-1}$ *to* $\overline{B(1)}$, *then problem* (VIII.16) *has at least one solution.*

Proof. We shall apply Theorem VII.4 to equation (VIII.17). Clearly, condition (a) of this theorem holds with $\delta = 0$. Now, taking Q to be the projector such that $\text{Im } Q = \text{span } (w_1', \ldots, w_d')$ and $\ker Q = \text{Im } L$, and using (VIII.18) and Lemma VIII.1, we see that

$$(\forall \text{ bounded } V \subset \text{Im } L)(\exists r_o > 0)(\forall r \geq r_o)(\forall v \in V)(\forall a \in S^{d-1}) :$$
$$QN(ra.w + v) \neq 0$$

which is essentially condition (b) of Theorem VII.4. Now, if

$0 < \mu \leq \inf_{a \in S^{d-1}} |\phi(a)| = \inf_{a \in S^{d-1}} |\tilde{\phi}(a)|$ and if $r_1 > 0$ is so large that

$$\sup_{a\in S^{d-1}} |\Phi(0,r_1a) - \phi(a)| < \mu ,$$

we have, by Rouché's theorem that if

$$J : \operatorname{Im} Q \to \ker L , \sum_{i=1}^{d} b_i w_i' \to \sum_{i=1}^{d} b_i w_i ,$$

then,

$$d[JQN|\ker L, B(r_1), 0] = d[\Phi(0,r_1 .),B(1),0] = d[\tilde{\phi},B(1),0] \neq 0$$

and the proof is complete.

Corollary VIII.3. *Assume that conditions of section* 7 *hold. Then if there exists a linear map* $T : (\operatorname{Im} L \cap \ker B)^{\perp} \to \ker L \cap \ker B$ *such that*

$$\int_{Tz>0} h_+ z + \int_{Tz<0} h_- z > 0 \tag{VIII.19}$$

for all $z \in (\operatorname{Im} L \cap \ker B)^{\perp}$, *problem* (VIII.16) *has at least one solution.*

Proof. We shall apply Theorem VIII.3 and show that (VIII.19) implies the nonvanishing of $d[\tilde{\phi},B(\ ), 0]$. Condition (VIII.19) implies that T must be one-to-one because if $\ker T \neq \{0\}$ and $z_0 \in \ker T$, then for any z such that

$$\tau = \int_{Tz>0} h_+ z_0 + \int_{Tz<0} h_- z_0 \neq 0$$

and any $a \in \underline{R}$ one has

$$\int_{T(z+az_o)>0} h_+ .(z+az_0) + \int_{T(z+az_o)<0} h_- .(z+az_0)$$

$$= a\tau + \int_{Tz>0} h_+ z + \int_{Tz<0} h_- z$$

which can be made negative by choosing suitably a. Thus if $w_1',\dots,w_d'$ is a basis in $(\operatorname{Im} L \cap \ker B)$ $(w_1,\dots,w_d)$ with $w_i = Tw_i'$ $(i = 1,\dots,d)$ will be a basis in $(\ker L \cap \ker B)$ and hence the mapping ϕ as used in Theorem VIII.3 can be taken as

$$\phi_i(a) = \int_{T(a.w')>0} h_+ w_i' + \int_{T(a.w')<0} h_- w_i'$$

and hence , using (VIII.19),

$$\sum_{i=1}^{d} a_i \phi_i(a) = \int_{T(a.w')>0} h_+(a.w') + \int_{T(a.w')<0} h_-(a.w') > 0$$

which implies, using Poincaré-Bohl theorem, that $d[\tilde{\phi}, B(1),0] = 1$, and achieves the proof.

12. Bibliographical notes concerning Chapter VIII.

The regularization results used in section 2 are due to Stampacchia (*Ann. Mat. Pura Appl.* (4) 51(1960) 1-37) and Fiorenza (*Ric. di Matem.* 14(1965) 102-123). Theorem VIII.1 is given in Mawhin ("Equations différentielles et fonctionnellles non linéaires, Hermann, Paris, 1973, 124-134), as well as Corollary VIII.1. Theorem VIII.2 is a slight improvement, with a quite different proof, of a result of de Figueireido (Some remarks on the Dirichlet problem for semilinear elliptic equations, Univ. Brasília, *Trabalho de Matemâtica* n° 57, March 1974) which proved it using a perturbation argument introduced by Hess (*Indiana J. Math.*, 23(1974) 827-830). For $\beta = 0$ and a second order linear part, Theorem VIII.2 and Corollary VIII.2 were first proved using Schauder fixed point theorem and an alternative argument by Landesman and Lazer (*J. Math. Mechanics* 19(1970) 609-623). This paper is really the source of all the material considered in this chapter, and has been extended in various ways, using the same technique of proof, by S.A. Williams (*J. Differential Equations* 8(1970) 580-586), J. Necas (*Comment. Math. Univ. Carolinae* 14(1973), 63-72), S. Fucik, M. Kucera and J. Necas (*J. Differential Equations* 17 (1975) 375-394). The results of section 7 to 11 are due to Nirenberg ("Trois. Coll CBRM Analyse fonctionnelle, Vander, Louvain, 1971, 57-74). See also Brezis ("Une équation elliptique nonlinéaire à la resonance", *Séminaire Lions-Schwartz*, 1971). The proof of Theorem VIII.3 is different from Nirenberg's one. See also Cronin (*J. Differential Equations* 14(1973) 581-596). Corollary VIII.3 is due to Schechter (*Ann. Scuola Norm. Sup. Pisa* (1973) 707-716 The proof given here is different.

It will be noted that the more general assumptions concerning the regularity of the coefficients of the linear part and concerning the growth restrictions on the nonlinear term are the ones of Theorem VIII.1. But this theorem only concerns a second order elliptic equation and the Neumann problem. Theorem VIII.2 is more general with respect to the order of the equation but the growth restriction on the nonlinear part is more restrictive and the linear part has to be selfadjoint. This condition is not necessary in Theorem VIII.3 which deals with fairly general linear part. with smooth coefficients, but requires the unique continuation property and a right hand member bounded everywhere.

IX. PERIODIC SOLUTIONS OF ORDINARY DIFFERENTIAL EQUATIONS WITH QUASIBOUNDED NONLINEARITIES AND OF FUNCTIONAL DIFFERENTIAL EQUATIONS

1. Let us study the existence of 2π-periodic solutions for the system

$$x'' + Dx = f(t,x) \qquad \text{(IX.1)}$$

where $f : R \times R^n \to R^n$ is 2π-periodic in t and continuous and $D = \operatorname{diag}(k_1^2,\dots,k_n^2)$ the numbers k_i $(1 \leqslant i \leqslant n)$ being integers. We shall apply to (IX.1) the theorem VII.5 and therefore we shall take $X = Z$ to be the Banach space P of mappings $x : R \to R^n$ which are continuous and 2 -periodic, with the norm $|x| = \sup_{t\in R} |x(t)|$, $|.|$ being say the Euclidian norm in R^n.

If dom L is the subspace of P of twice continuously differentiable 2π-periodic mappings and if

$$L : \operatorname{dom} L \subset P \to P, \; x \to x'' + Dx$$

$$F : P \to P \;,\; x \to f(.,x(.)) \;,$$

then F is clearly continuous and the 2π-periodic solutions of (IX.1) are the solutions x of

$$Lx = Fx \;.$$

Now

$$\ker L = \{x : x_i(t) = a_i \cos k_i t + b_i \sin k_i t,\; a_i, b_i \in R,\; 1 \leqslant i \leqslant n\}$$

$$\operatorname{Im} L = \{y \in P : \int_0^{2\pi} y_i(t) \cos k_i + dt = \int_0^{2\pi} y_i(t) \sin k_i t \, dt = 0, \quad 1 \leqslant i \leqslant n\}$$

and it is clear that the projectors $P, Q : P \to P$ defined by

$$(Px)_i(t) = (Qx)_i(t) =$$

$$= \pi^{-1} \cos k_i t \int_0^{2\pi} x_i(t) \cos k_i t \, dt + \pi^{-1} \sin k_i t \int_0^{2\pi} x_i(t) \sin k_i t \, dt \qquad 1 \leqslant i \leqslant n$$

are such that

$$\operatorname{Im} P = \ker L \;,\; \operatorname{Im} L = \ker Q$$

and are orthogonal in the Hilbert space $U = L_n^2(0,2\pi)$ of measurable and square integrable mappings $x : [0,2\pi] \to R^n$ with the inner product

$$(x,y) = \int_0^{2\pi} \langle x(t), y(t) \rangle \, dt$$

where $< u,v > = \sum_{i=1}^{n} u_i v_i$ is the inner product in R^n.

Theorem IX.1. *The system* (IX.1) *has at least a* 2π*-periodic solution provided that*

(i) $(\exists\, \alpha \in [0,1[)$, $\lim_{|x| \to \infty} |x|^{-\alpha} |f(t,x)| = 0$

uniformly in t.

(ii) *there exists a continuous* 2π*-periodic matrix function* $K : R \to L(R^n, R^n)$ *and a matrix* $\tilde{H} = \operatorname{diag}(\alpha_1, \ldots, \alpha_n)$ *with* $\alpha_i = \pm 1$ *such that*

a) $K(t)\, x(t) = 0$ *on* R *and* $x \in \ker L$ *imply* $x = 0$.

b) $(\exists\, \beta \in [0,\alpha[,\ 2\alpha - 1 < \beta)$ $(\exists\, a > 0,\ \rho \geq 0)$ $(\forall t \in R)$ $(\forall x \in R^n : |x| \geq \rho$

$< \tilde{H}x, f(t,x) > \ \geq a\, |K(t)\, x|^{1+\beta}$.

Proof. Let us take, in the notations of Theorem VII.5 the space X,Z,U as defined above and $V = L_n^{1+\beta}(0,2\pi)$ to be space of measurable mappings $x : [0,2\pi] \to R^n$ such that $|x|^{1+\beta}$ is integrable with the usual norm. The imbeddings conditions are satisfied for those spaces. Now condition (i) implies that numbers $\mu', \nu' \geq 0$ exist such that, for all $t \in R$ and $x \in R^n$,

$$|f(t,x)| \leq \mu' |x|^{\alpha} + \nu'$$

so that for all $x \in P$,

$$|Fx| \leq \mu' |x|^{\alpha} + \nu'.$$

The right inverse $K_{P,Q}$ of L being continuous by the closed graph theorem (L being obviously closed) we also have

$$|K_{P,Q} Fx| \leq \mu\, |x|^{\alpha} + \nu$$

for some $\mu \geq 0$, $\nu \geq 0$. Now the operators $H,G : P \to P$ defined by

$$(Hx)(t) = \tilde{H}x(t)$$

$$(Gx)(t) = K(t)\, x(t)$$

for $t \in R$ are clearly such that $H \mid \ker L$ is an isomorphism between $\ker L$ and $\operatorname{Im} Q$ ($= \ker L$) and $G \mid \ker L$ is injective. Moreover there exists $b > 0$ such that

$$(\forall t \in R)\ (\forall x \in R^n) : < \tilde{H}x, f(t,x) > \ \geq a\, |K(t)x|^{1+\beta} - b$$

which implies that, for all $x \in P$

$$(Hx, Fx) \geq a\, \|Gx\|^{1+\beta} - b$$

where $\|.\|$ is the norm of $L_n^{1+\beta}(0,2\pi)$. Thus all the assumptions of Theorem IX.1 are satisfied and the proof is complete.

Corollary IX.1. *The system* (IX.1) *has at least a* 2π*-periodic solution provided that condition* (i) *of Theorem* IX.1 *holds and*

(ii') *There exists a matrix* $\tilde{H} = \text{diag}(\alpha_1,\ldots,\alpha_n)$ *with* $\alpha_i = \pm 1$ *and numbers* $\beta < \alpha$, $a > 0$, $\rho \geq 0$ *such that* $\beta \geq 0$, $\beta > 2\alpha - 1$ *and*

a) $(\forall t \in R)\ (\forall x \in R^n : |x| \geq \rho) : \langle Hx, f(t,x)\rangle \geq 0.$

c) $(\exists t_i \in [0,2\pi[\,,\ 1 \leq i \leq n)\ (\exists\ \varepsilon > 0)\ (\forall\ \tau \in [t_i - \varepsilon, t_i + \varepsilon])$
$(\forall x \in R^n : |x| \geq \rho) : \langle \tilde{H}x, f(\tau,x)\rangle \geq a\,|x_i|^{1+\beta}.$

Proof. The corollary corresponds to the choice

$$K(t) = (K_{ij}(t))$$

with $K_{ij}(t)$ 2π-periodic and continuous, $K_{ij}(t) = \delta_{ij}\,\delta_{ik}$ for $t \in [t_k - \varepsilon, t_k + \varepsilon]$ and $K_{ij}(t) = 0$ for $t \in [0,2\pi[\ \setminus [t_k - 2\varepsilon, t_k + 2\varepsilon]$.

A particular case of interest is the scalar equation

$$x'' + n^2 x = f(t,x) \qquad \text{(IX.2)}$$

for which the following result holds from Corollary IX.1 with $\beta = 0$

Corollary IX.2. *The equation* (IX.2) *has at least one* 2π*-periodic solution provided that*

(i) $(\exists\ \alpha \in [0,\tfrac{1}{2}[) : \lim_{|x|\to\infty} |x|^{-\alpha}\,|f(t,x)| = 0$ *uniformly in* t.

(ii) $(\exists\ \rho \geq 0)\ (\exists\ a > 0)$ such that with $i = 1$ or -1,

a) $(\forall t \in R)\ (\forall x \in R : |x| \geq \rho) : i \text{ sign } xf(t,x) \geq 0;$

b) $(\exists t^* \in [0,2\pi[)\ (\exists \varepsilon > 0)\ (\forall \tau \in [t^* - \varepsilon, t^* + \varepsilon])$
$(\forall x \in R : |x| \geq \rho) : i \text{ sign } xf(t,x) \geq a.$

Consider now the scalar equation

$$x'' + n^2 x = f(t,x) + h(t) \qquad \text{(IX.3)}$$

with $h : R \to R$ continuous and 2π-periodic.

Theorem IX.2. *Equation* (IX.3) *has at least a* 2π*-periodic solution provided that*

(i) $(\exists \alpha \in [0,\tfrac{1}{2}[) : \lim_{|x|\to\infty} |x|^{-\alpha} |f(t,x)| = 0$ *uniformly in* t ;

(ii) $(\exists B \geq 0)\ (\exists \rho \geq 0)\ (\forall t \in R)\ (\forall x \in R : |x| \geq \rho)$:
$i \text{ sign } x\, f(t,x) \geq B$

(iii) $4B > \pi A$,

where $i = -1$ or $+1$, $A = (a^2 + b^2)^{\frac{1}{2}}$,

$$a = \pi^{-1} \int_0^{2\pi} h(t) \cos nt \, dt, \quad b = \pi^{-1} \int_0^{2\pi} h(t) \sin nt \, dt.$$

Proof. We shall apply Theorem VII.5 to the corresponding abstract equation

$$Lx - Fx = h$$

where $Lx = x'' + n^2 x$, $Fx = f(.,x(.))$ and for which the needed regularity assumptions hold as shown previously. We also choose $\beta = 0$, $Hx = ix$, $Gx = x$. It follows from (i) that constants $\mu, \mu', \nu, \nu' \geqslant 0$ exist so that

$$|K_{P,Q} Fx| \leqslant \mu |x|^{\alpha} + \nu, \quad |Fx| \leqslant \mu' |x|^{\alpha} + \nu'$$

for all $x \in P$. It follows from (ii) that a constant b exists such that

$$i \, xf(t,x) \geqslant B|x| - b$$

for $t \in \underline{R}$ and $x \in R$ so that

$$(Hx, Fx) > B \|x\| - 2\pi b = B \|Gx\| - 2\pi b$$

for all $x \in P$. Therefore the existence will follow from Theorem VII.5 under the condition (VII.14), i.e.

$$\sup_{w \in \ker L \cap \partial B(1)} \|w\|^{-1} (iw, Ph) < B$$

But if $w(t) = c \cos nt + d \sin nt$, $c^2 + d^2 = 1$,

$$\|w\|^{-1} (iw, Ph) =$$

$$= i \frac{\int_0^{2\pi} (c \cos nt + d \sin nt)(a \cos nt + b \sin nt) \, dt}{\int_0^{2\pi} |c \cos nt + d \sin nt| \, dt}$$

$$= \frac{i\pi(ac + bd)}{\int_0^{2\pi} |\sin(nt + \phi)| \, dt} \qquad (\sin \phi = c, \; \cos \phi = d)$$

$$= i \frac{\pi}{4} (ac + bd) \qquad (i = \pm 1).$$

$$= i \frac{\pi}{4} \langle \gamma \, \delta \rangle \quad \text{where} \quad \gamma = (a,b), \quad \delta = (c,d)$$

Obviously

$$\sup_{c^2+d^2=1} i\,\frac{\pi}{4} < \gamma,\delta > \;=\; \frac{\pi}{4}\,(a^2+b^2)^{\frac{1}{2}} = \frac{\pi A}{4}$$

and the proof is complete.

2. If $l \geq 0$ is an integer, let us now denote by P_T^l the (Banach) space of continuous and T-periodic mappings $x : R \to R^n$ with the norm

$$|x|_l = \sum_{j=0}^{l} [\sup_{t\in R} |x^{(j)}(t)|]$$

where $x^{(j)} = d^jx/dt^j$ and $|.|$ is the Euclidian norm in R^n. Let us introduce the projector

$$P : P_T \to P_T\,,\; x \to T^{-1}\int_0^T x(t)\,dt\,.$$

(We shall write P_T for P_T^0 and $|.|$ for $|.|_0$.).

It is immediate that for every $x \in P_T$,

$$|Px| \leq |x|$$

and that Im P is the subspace of P_T of constant mappings.

If $k \geq 0$ is an integer, let us summarize some properties of the vector differential operator with constant coefficients L defined by

$$Lx = x^{(k+1)} + A_1x^{(k)} + \ldots + A_kx' + A_{k+1}x \qquad \text{(IX.4)}$$

where the $A_i (i = 1,\ldots,k+1)$ are $(n \times n)$ constant matrices and

$$\text{dom } L = \{x\in P_T : x^{(k+1)} \text{ exists and is continuous}\}.$$

It is then clear that

$$\text{Im } L \subset P_T\,.$$

It is well known that the adjoint L^* of L is the operator defined by

$$L^*u = u^{(k+1)} - u^{(k)}(.)A_1 + \ldots + (-1)^k u'(.)A_k + (-1)^{k+1}u(.)A_{k+1},$$

where $u : R \to (R^n)^*$ is T-periodic and has continuous derivatives up to the order k+1 ($(R^n)^*$ is the dual space of R^n).

The following result is classical and we recall it only for completeness. I_n will denote the $(n \times n)$ identity matrix and $\omega = 2\,/T$.

<u>Proposition IX.1</u>. *If* L *is defined by* (IX.4), ker L $\neq$ {0} *if and only if the equation*

$$\text{dét}(\lambda^{k+1}I_n + \lambda^kA_1 + \ldots + A_{k+1}) = 0 \qquad \text{(IX.5)}$$

has roots of the form $\lambda = im\omega$, *with* m *an integer.* ker L *and* ker L^* *have the same dimension and* ker L (*resp.* ker L^*) *is formed by the elements of* dom L (*resp.* dom L^*) *obtained by taking the real and the imaginary parts of the complex mappings*

$$t \to \exp(im\omega t)c \quad (\textit{resp. } t \to \exp(im\omega t)d\ ),$$

with $im\omega$ *a root of* IX.5 *and* c *the column n-vectors (resp.* d *the row n-vectors) formed by the* n *first components of the generalized eigenvectors, relative to* $im\omega$, *of the* $[(k+1)n \times (k+1)n]$ *matrix*

$$A = \begin{pmatrix} 0_n & I_n & 0_n & \cdots & 0_n \\ & & & & \\ \cdot & & & & \\ 0_n & 0_n & \cdots & 0_n & I_n \\ -A_{k+1} & -A_k & \cdots & -A_2 & -A_1 \end{pmatrix}$$

(*resp. the* n *last components of the generalized eigenvectors, relative to* $-im\omega$, *of minus the transposed of* A). *Lastly,*

$$\operatorname{Im} L = \{x \in P_T^0 : \int_0^T u(t)x(t)dt = 0\ , \quad u \in \ker L^*\} \tag{IX.6}$$

(Fredholm alternative) and there exists a constant $\kappa \geq 0$ *such that*

$$|Kx|_k \leq \kappa |x| \tag{IX.7}$$

for every $x \in \operatorname{Im} L$, *with* K *the (unique) right inverse of* L *taking values in a fixed topological supplement of* $\ker L$ *in* P_T.

The following corollary will be particularly useful in the sequel.

Proposition IX.2. $\ker L = \{x \in \operatorname{dom} L : x$ *is a constant mapping*$\} = \operatorname{Im} P$ (IX.8) *if and only if*

$$A_{k+1} = 0_n \tag{IX.9}$$

and equation

$$\det(\lambda^k I_n + \lambda^{k-1} A_1 + \ldots + A_k) = 0 \tag{IX.10}$$

has no root λ *of the form* $im\omega$ *a nonzero integer. In this case*

$$\operatorname{Im} L = \{x \in P_T : Px = 0\} \tag{IX.11}$$

and the unique right inverse K_P *of* L *such that* $PK = 0$ *is compact.*

Proof. 1. Necessity. ker L being the subset of dom L of constant mappings, we have

$$A_{k+1}c = 0$$

for every $c \in R^n$ and hence IX.9 is satisfied. Now, by IX.8 and Proposition IX.1, the equation

$$\text{dét}(\lambda^{k+1}I_n + \lambda^k A_1 + \ldots + \lambda A_k) \equiv \lambda^n \text{dét}(\lambda^k I_n + \lambda^{k-1}A_1 + \ldots + A_k) = 0 \tag{IX.12}$$

has no solution of the form $im\omega$ with m a nonzero integer, and the same is true for equation (IX.10). $\ker L^*$ being of dimension n and containing the set of constant mappings from R into $(R^n)^*$ coincides with it and then, by taking in IX.6 for u successively the mappings

$$t \to e_i^* = (0,\ldots,0,1,0,\ldots 0) \quad (i = 1,2,\ldots,n) ,$$

we obtain (IX.11). To prove the compactness of K_P let B be any bounded set in Im L; thus there exists $b \geq 0$ such that

$$|v| \leq b , \quad v \in B,$$

and hence, by (4.4),

$$|K_P v| \leq \kappa b , \quad v \in B,$$

which shows that the set

$$B' = \{K_P v : v \in B\}$$

is bounded in P_T^k. On the other hand, every $K_P v \in \text{dom } L$, and thus has continuous derivatives up to the order k+1, and verifies the relation

$$(Kv)^{(k+1)} + A_1(Kv)^{(k)} + \ldots + A_k(Kv)' = v ,$$

which implies that

$$\sup_{t\in R} |(Kv)^{(k+1)}(t)| \leq \sum_{j=1}^{k} |A_j| \kappa b + b .$$

Then, by a direct argument using Arzela-Ascoli theorem, B' is relatively compact and hence K is a compact mapping.

2. Sufficiency. By (IX.9), ker L contains the subset of dom L of constant mappings and, using (IX.12) and the condition upon the roots of (IX.10) ker L can only be spanned by constant mappings and is then necessarily equal to (IX.8). The remaining of the proof is the same than that of necessity.

3. We shall now be interested in the study of the T-periodic solutions of the vector ordinary differential equation

$$x^{(k+1)} + A_1 x^{(k)} + \ldots + A_k x' = f(t,x) \tag{IX.13}$$

where $f : R \times R^n \to R^n$ is T-periodic in t and continuous and such that there exists $\delta \in [0,1[$, $\mu',\nu' \geq 0$ such that, for all $t \in R$ and $x \in R^n$,

$$|f(t,x)| \leq \mu'|x|^\delta + \nu' \tag{IX.14}$$

with $|\cdot|$ the Euclidian norm in R^n.

Theorem IX.3. *Assume that the following conditions hold :*

(i) *Equation* (IX.10) *has no* root λ *of the form* $im\omega$ *with* m *a nonzero integer.*

(ii) *There exists a matrix* $E = \text{diag}(\varepsilon_1,\dots,\varepsilon_n)$ *with* $\varepsilon_i = \pm 1$ $(i = 1,\dots,n)$ *such that*

$$\lim_{|x|\to\infty} \frac{< Ex, f(t,x) >}{|x|^{2\delta}} = \infty$$

uniformly in $t \in R$, *with* δ *given in* (IX.14) .

Then equation (IX.13) *has at least one* T*-periodic solution*

Proof. We shall apply Theorem VII.4 to the equivalent abstract equation in P_T

$$Lx = Nx$$

where L is defined in (IX.4) (with $A_{k+1} = 0$) and

$$(Nx)(t) = f(t,x(t)).$$

Clearly N is a continuous mapping from P_T into itself and it follows from (IX.14) and Proposition IX.1 that, for all $x \in P_T$,

$$|K_{P,Q}\, Nx| \leqslant \mu |x|^{\delta} + \nu$$

for some $\mu,\nu \geqslant 0$, and that N is L-compact on bounded sets. Also L is Fredholm of index zero and

$$\ker L = \operatorname{Im} P \ , \ \operatorname{Im} L = \ker P$$

with P defined in the beginning of section 1. Thus condition (a) of Theorem VII.4 is satisfied. If condition (b) of the same theorem is not verified, then

$$(\exists \text{ bounded } V \subset \operatorname{Im} L)\ (\forall t_0 > 0)\ (\exists\, t \geqslant t_0)\ (\exists\, z \in V)$$

$$(\exists\, w \in \ker L \cap \partial B(1)) : \int_0^T f(\tau, tw + t^{\delta} z(\tau))\, d\tau = 0$$

and hence

$$(\exists \text{ bounded } V \subset \operatorname{Im} L)\ (\exists \text{ sequence } (t_n) \text{ with } t_n > 0,\ t_n \to \infty \text{ if } n \to \infty)$$

$$(\exists \text{ sequence } (z_n),\ z_n \in V)\ (\exists \text{ sequence } (w_n),\ w_n \in \ker L \cap \partial B(1)) :$$

$$\int_0^T \frac{< Et_n w_n, f(\tau, t_n w_n + t_n z_n(\tau)) >}{t_n^{2\delta}}\, d\tau = 0 \qquad \text{(IX.15)}$$

Now, for all $\tau \in [0,T]$, using the fact that $z_n \in V$ for all n,

$$|w_n + t_n^{\delta-1} z_n(\tau)| \geq 1 - t_n^{\delta-1} C \geq \frac{1}{2}$$

if $n \geq N$, N large enough and hence

$$|t_n w_n + t_n^{\delta}.z_n(\tau)| \geq \frac{t_n}{2}, \; n \geq N, \; \tau \in [0,T] . \qquad \text{(IX.16)}$$

(IX.15) is clearly equivalent to

$$\int_0^T \frac{< E(t_n w_n + t_n^{\delta} z_n(\tau)), f(\tau, t_n w_n + t_n^{\delta} z_n(\tau)) >}{t_n^{2\delta}} d\tau =$$

$$\int_0^T \frac{< E(t_n^{\delta} z_n(\tau)), f(\tau, t_n w_n + t_n^{\delta} z_n(\tau)) >}{t_n^{2\delta}} d\tau$$

which implies, for $n \geq N$,

$$\frac{1}{2^{2\delta}} \int_0^T \frac{< E(t_n w_n + t_n^{\delta} z_n(\tau)), f(\tau, t_n^{\delta} w_n + t_n z_n(\tau) >}{|t_n w_n + t_n^{\delta} z_n(\tau)|^{2\delta}} d\tau$$

$$\leq \int_0^T C t_n^{-\delta}(\mu' |t_n w_n + t_n^{\delta} z_n(\tau)|^{\delta} + \nu') d\tau$$

$$\leq \int_0^T C(\mu' |w_n + t_n^{\delta-1} z_n(\tau)|^{\delta} + t_n^{-\delta}\nu') d\tau$$

$$\leq D = D(C,\mu',\nu') < \infty . \qquad \text{(IX.17)}$$

Now by assumption (ii) ,

$$(\exists\, \rho > 0)\ (\forall x \in P_T^0 : \inf_{\tau \in [0,T]} |x(\tau)| \geq \rho :$$

$$\int_0^T \frac{< Ex(\tau), f(\tau, x(\tau)) >}{|x(\tau)|} d\tau > 2^{2\delta} D$$

and by (IX.16)

$$(\exists\, N' \geq N)\ (\forall n \geq N') : \inf_{\tau \in [0,T]} |t_n w_n + t_n^{\delta} z_n(\tau)| \geq \rho$$

which leads to a contradiction to (IX.17) when $n \geq N'$.

Lastly, if $c \in R^n$ is such that $|c|$ is large enough, we deduce from assumption (ii) that

$$< Ec, \frac{1}{T} \int_0^T f(t,c)dt > \; > 0$$

which implies by the Poincaré - Bohl theorem that

$$d\,[\,PN|\ker L,\ B(\ ),\ 0] = \pm 1$$

for τ large enough and achieves the proof.

4. For some $r \geq 0$, let $\mathcal{C}_r$ be the Banach space of continuous mappings $\phi : [-r,0] \to R^n$ with the norm

$$\|\phi\| = \sup_{\theta\in[-r,0]} |\phi(\theta)| .$$

When $r = 0$, $\mathcal{C}_r$ is naturally identified to R^n.

If $x \in P_T$ and $t \in R$, we shall denote by x_t the element of $\mathcal{C}_r$ defined by

$$x_t : [-r,0] \to R^n,\ \theta \to x(t+\theta).$$

We note that,

$$|x_t| = \sup_{\theta\in[-r,0]} |x(t+\theta)| \leq \sup_{t\in R} |x(t)| = |x| .$$

When $r = 0$, the mapping x_t will be naturally identified with the element $x(t)$ of R^n. Moreover we shall sometimes identify, without further comment, a constant mapping in P_T or $\mathcal{C}_r$ with the element of R^n given by its constant value.

Let Lx be given by (IX.4) with the assumptions of Proposition IX.2 and let

$$f : R \times \mathcal{C}_r \to R^n,\quad (t,\phi) \to f(t,\phi)$$

be T-periodic with respect to t, continuous and takes bounded sets into bounded sets. Let us consider the functional differential equation

$$Lx \equiv x^{(k+1)} + A_1 x^{(k)} + \dots + A_k x' =$$

$$= f(t, x_t). \qquad \text{(IX.18)}$$

Thus, up to the end of the Chapter, the roots λ of (IX.10) will not be of the form $im\omega$ with m a nonzero integer.

If we define

$$N : P_T \to P_T,\quad x \to f(.,x_.)$$

then it is clear that the T-periodic solutions of (IX.18) are the solutions in P_T of

$$Lx = Nx \qquad \text{(IX.19)}$$

and we have shown that L is a Fredholm mapping of index zero, that

$$\ker L = \operatorname{Im} P,\ \operatorname{Im} L = \ker P$$

and that $K_{P,P}$ is compact. As N clearly takes bounded sets into bounded sets,

$K_{P,P}N$ will take bounded sets into relatively compact sets.

Lemma IX.1. *With the assumptions and notations above,* N *is* L*-compact on each bounded set of* P_T .

Proof. We have already shown that $K_{P,P}N$ takes bounded sets into relatively compact sets and the same is trivially true for ΠN . Let us show that $K_{P,P}N$ is continuous. Let (x^n) be a sequence which converges in P_T to x. Then $\{x^n, n = 1, 2, \ldots\}$ is bounded and $\{K_{P,P}Nx^n , n = 1, 2, \ldots\}$ belongs to a compact set of P_T . Hence, there exist a subsequence (x^{n_j}) of (x^n) such that $K_{P,P}Nx^{n_j} \to y$ if $j \to \infty$. For each $s \in R$, $x^n_s \to x_s$ in $\mathcal{C}_r$ and hence, by the continuity of f,

$$f(s, x^n_s) \to f(s,x_s)$$

as $n \to \infty$. On the other hand, because f takes bounded sets into bounded sets and because of its T-periodicity in t, there exists $M > 0$ such that, for

$$|f(s, x^n_s)| \leq M .$$

Hence, using the Lebesgue convergence theorem, we obtain that, for each $t \in R$,

$$y(t) = \lim_{j\to\infty} (K_{P,P}Nx^{n_j}_s)\ (t) = \lim_{j\to\infty} \int_0^t \left[f(s,x^{n_j}_s) - T^{-1} \int_0^T f(u,x^{n_j}_s)du \right] ds$$

$$= \int_0^t \left[f(s,x_s) - T^{-1} \int_0^T f(u,x_u)du \right] ds = (K_{P,P}Nx)\ (t) .$$

Thus, the limit of any convergent subsequence of (x^n) is independent of the subsequence. Every subsequence of $(K_{P,P}Nx^n)$ having a convergent subsequence with the limit $K_{P,P}Nx$ independent of this subsequence, a classical argument shows that $(K_{P,Q}Nx^n)$ itself converges to $K_{P,Q}Nx$, and $K_{P,Q}N$ is continuous. The same argument can be used to show that QN, and hence ΠN , is continuous and the proof is complete.

5. Before considering quasibounded nonlinearities let us give for functional differential equations a few results in the line of Chapter V.

Definition IX.1. A $\mathcal{C}^1$ function $V : R^n \to R$, $x \to V(x)$ will be said to be a *guiding function* for equation (IX.18) with $k = 0$ if there exists $\rho > 0$ such that

$$< \text{grad } V(x(t)), f(t, x_t) >> 0 \qquad \text{(IX.20)}$$

for every $x \in P_T$ and every $t \in R$ for which

$$|x(t)| \geq \rho \text{ and } |V(x(t))| \geq |V(x(s))| ,\ s \in [0,T] .$$

In particular if (IX.18) is an ordinary differential equation, a sufficient condition for V to be a guiding function is that, for $|x| \geq \rho$, $x \in R^n$, and $t \in R$,

$$< \text{grad } V(x), f(t, x) >> 0 ,$$

which is the condition given in Chapter V.

Also, if

$$h : R \times R^n \times \mathscr{C}_r \to R^n , \quad (t,x,\phi) \to h(t,x,\phi)$$

is T-periodic with respect to t, continuous and takes bounded sets into bounded sets, and if there exists $\rho > 0$ such that, for each $t \in R$, each $\phi \in \mathscr{C}_r$ and each $|x| \geq \rho$ we have

$$< \text{grad } V(x), h(t,x,\phi) >> 0 ,$$

then V is a guiding function for the functional differential equation

$$x'(t) = h(t,x(t),x_t). \qquad (IX.21)$$

When we shall consider a family $\{V_j\}$ of guiding functions we shall write ρ_j for the corresponding value of ρ.

Theorem IX.4. *If there exist* $m + 1$ (≥ 1) *guiding functions* $V_0, V_1, \ldots, V_m$ *for* (IX.18) with $k = 0$ *such that*

$$\lim_{|x| \to \infty} [|V_0(x)| + |V_1(x)| + \ldots + |V_m(x)|] = \infty \qquad (IX.22)$$

and

$$d [\text{grad } V_0, B(\rho_0), 0] \neq 0$$

then (IX.18) with $k = 0$ *has at least one T-periodic solution.*

Proof. We will apply Corollary IV.1 to (IX.19) with $N^*x = Nx$. Let $\lambda \in]0,1[$ and x be a possible solution of

$$Lx = \lambda Nx ,$$

i.e. a possible T-periodic solution of equation

$$x' = \lambda f(t,x_t). \qquad (IX.23)$$

Let us write

$$\mathcal{V}_j(t) = V_j[x(t)] , \; j = 0,1,\ldots,m ; \quad t \in R .$$

Then, $\mathcal{V}_j$ is of class $\mathscr{C}^1$ and

$$\mathcal{V}_j'(t) = < \text{grad } V_j \; x(t) , x'(t) >$$

$$= \langle \operatorname{grad} V_j[x(t)], f(t,x_t) \rangle, \; j = 0,1,\ldots,m; \; t \in R.$$

For every τ_j such that

$$V_j(\tau_j) = \sup_{t \in R} V_j(t) = \sup_{t \in [0,T]} V_j(t), \; j=0,1,\ldots,m,$$

i.e. such that

$$|V_j[x(t)]| \leq |V_j[x(\tau_j)]|, \; t \in R,$$

$V_j(t)$ is a maximum or minimum and hence

$$V_j'(\tau_j) = 0, \; j = 0,1,\ldots,m.$$

Then, using (IX.20),

$$|x(\tau_j)| < \rho_j, \; j = 0,1,\ldots,m,$$

and, for each $t \in R$,

$$|V[x(t)]| \leq |V[x(\tau_j)]| \leq \sup_{|y| \leq \rho_j} |V(y)| = M_j,$$

which implies that

$$\sum_{j=0}^{m} |V_j[x(t)]| \leq \sum_{j=0}^{m} M_j = M, \; t \in R. \tag{IX.24}$$

Using (IX.22), there exists $\rho_M > 0$ such that

$$\sum_{j=0}^{m} |V_j(y)| > M$$

if $|y| \geq \rho_M$, and then (IX.24) implies that

$$\sup_{t \in [0,T]} |x(t)| < \rho_M$$

for each possible T-periodic solution of (IX.23).

Now, using (VIII.4) with $x(t) = a$, a constant, we obtain

$$\langle \operatorname{grad} V_0(a), \int_0^T f(t,a)dt \rangle > 0$$

for every $a \in R^n$ such that $|a| \geq \rho_0$ and hence each possible solution of equation

$$PN(a) \equiv T^{-1} \int_0^T f(t,a)dt = 0$$

is in $B(\rho_0)$ and, using Poincaré-Bohl theorem,

$$d[PN| \ker L, B(\rho), 0] = d[\operatorname{grad} V_0, B(\rho), 0] \neq 0$$

for each $\rho \geq \rho_o$. All the conditions of Corollary IV.1 are satisfied and the proof is complete.

Example. Let us consider the scalar delay-differential equation

$$x'(t) = ax(t) + bx(t-r) + cx^3(t) + dx^3(t-r) + e(t) = f(t,x_t) \qquad \text{(IX.25)}$$

where a,b,c,d are constant with

$$|c| > |d|$$

and $e : R \to R$ is continuous and T-periodic.

Let

$$V(x) = (1/2)cx^2 .$$

Then, if x(t) is T-periodic,

$$V' x(t) f(t,x_t) = acx^2(t) + bcx(t)x(t-r) + c^2x^4(t) + cdx^3(t-r)x(t) + ce(t)x(t)$$

$$\geq c^2|x(t)|^4 - |c||d||x(t-r)|^3|x(t)| - |a||c||x(t)|^2$$

$$- |b||c||x(t)||x(t-r)| - |c||e(t)||x(t)| .$$

For t such that $|x(t)| \geq \rho$ and

$$(1/2)|c||x(t)|^2 \geq (1/2)|c||x(s)|^2 , \quad s \in R ,$$

i.e.

$$|x(t)| \geq \rho , \ |x(t)| \geq |x(s)| , \quad s \in R ,$$

we have, because $|x(t-r)| \leq |x(t)|$,

$$V'|x(t)| f(t,x_t) \geq c^2|x(t)|^4 - |c||d||x(t)|^4 - |a||c||x(t)|^2 -$$

$$- |b||c||x(t)|^2 - |c| \sup_{s\in R}|e(s)||x(t)|$$

$$= |c||x(t)|^4\left[|c| - |d| - (|a| + |b|)|x(t)^{-2} - \sup_{s\in R}|e(s)||x(t)|^{-3}\right]$$

$$\geq |c|\rho^4\left[|c| - |d| - (|a| + |b|)\rho^{-2} - \rho^{-3}\sup_{s\in R}|e(s)|\right] > 0 \text{ if } \rho \text{ is}$$

sufficiently great.

All the conditions of Theorem IX.4 are satisfied for the guiding function $V(x) = (1/2)cx^2$ and then IX.25 has one T-periodic solution.

6. We shall now introduce some geometrical conditions for the existence of periodic solutions. From the considerations of Chapter V recall that if $G \subset R^n$ is an open bounded convex set containing the origin, then, for each $x \in \partial G$, there exists at least one non-zero $n(x) \in R^n$ such that

(i) $n(x).x \neq 0$

(ii) $\overline{\Omega} \subset \{y : [< n(x),x >/|< n(x),x >|]\ n(x).(y - x) \leq 0\}$

Such a $n(x)$ has been called a *normal* to G at x'. If $n(x).x > 0$, it is an *outer normal*, if $n(x).x < 0$, an *inner normal.*

Theorem IX.5. *Let* $G \subset R^n$ *be an open bounded convex set containing the origin and suppose that there exists a normal* $n(x)$ *to* G *such that, for each* $x \in P_T$ *for each* $x \in P_T$ *for which* $x(t) \in \overline{G}$, $t \in R$ *and* $x(t_o) \in \partial G$ *for some* t_o, then

$$< n[x(t_o)],f(t_o, x_{t_o}) >> 0 . \qquad \text{(IX.26)}$$

for all such t_o .

Then, if

$$d[g, G, 0] \neq 0$$

with

$$g : R^n \to R^n , \; a \to T^{-1} \int_0^T f(t,a)dt$$

equation (IX.18) with $k = 0$ *has at least one T-periodic solution* x *such that* $x(t) \in G$, $t \in R$.

Proof. We apply Corollary IV.1 to (IX.19) with

$$\Omega = \{x \in P_T\ x(t) \in G,\ t \in R\} .$$

Hence

$$\partial\Omega = \{x \in P_T\ x(t) \in \overline{G} ,\ t \in R,\ x(t_o) \in \partial G \text{ for some } t_o\} .$$

Let $\lambda \in]0,1]$ and let x be a possible solution of

$$Lx = \lambda Nx . \qquad \text{(IX.27)}$$

If $x \in \partial\Omega$, then $x(t) \in \overline{G}$, $t \in R$ and, for some t_o, $x(t_o) \in \partial\Omega$. Hence, using condition (ii) in the definition of a normal we get, for each h,

$$\{[< n[x(t_o)],x(t_o) >]/|< n[x(t_o)],x(t_o) >|\} < n[x(t_o)],[x(t_o+h)-x(t_o)] > \leq 0$$

which implies that

$$< n[x(t_o)],x'(t_o) > = 0 .$$

But then, using (IX.26)

$$0 = < n[x(t_o)],x'(t_o)> = \lambda < n[x(t_o)],f(t_o,x_{t_o}) >> 0 ,$$

a contradiction. On the other hand, using (IX.26) with a constant mapping a, we have

$$< n(a), f(t, a) > > 0 \ , \ a \in \partial G$$

for each $t \in R$ and hence

$$< n(a), g(a) > > 0 \ , \ a \in \partial G$$

which shows that g cannot vanish on ∂G and achieves the proof.

<u>Corollary IX.3</u>. *If assumption* (IX.26) *holds with* $>$ *or* $<$ *and* n *an outer normal, then equation* (IX.18) *with* $k = 0$ *has at least one T-periodic solution* x *such that* $x(t) \in G$, $t \in R$.

<u>Proof</u>. We only have to show that in this case $d[g, G, 0] \neq 0$. For each $\lambda \in [0,1]$ and $a \in \partial G$, we have

$$\pm(1 - \lambda)n(a).a + \lambda n(a)T^{-1} \int_0^T f(t,a)dt \gtrless 0$$

with $+$ or $-$ according to (IX.26) holds with $>$ or $<$.

Hence

$$\pm(1 - \lambda)a + \lambda g(a) \neq 0 \ , \ a \in \partial G$$

and

$$d[g, G, 0] = d[\pm I, G, 0] = (\ 1)^n .$$

<u>Corollary IX.4</u>. *Let* G *be an open bounded convex set in* R^n *containing zero and assume that there exists an outer normal* $n(x)$ *to* ∂G *such that, for each* $t \in \partial G$ *and each continuous mapping* $y : [-r,0] \to \overline{G}$, *one has*

$$< n(x), h(t,x,y) > > 0 \qquad \text{(IX.28)}$$

(or < 0*). Then equation* (IX.21) *has at least one T-periodic solution* x *such that* $x(t) \in G$ *for each* $t \in R$.

<u>Proof</u>. If we write

$$f(t,x_t) = h(t,x(t),x_t),$$

if $x \in P_T$ is such that $x(t) \in \overline{G}$ for each $t \in R$ and if t_0 is such that $x(t_0) \in \partial G$, then $x_{t_0}(\theta) = x(t_0 + \theta) \in \overline{G}$ for each $\theta \in [-r,0]$ and hence

$$< n[x(t_0)], f(t_0, x_{t_0}) > = < n[x(t_0)], h(t_0, x(t_0), x_{t_0}) > > 0$$

(or < 0) by (IX.28) which implies that Corollary IX.3 can be applied.

<u>Remark</u>. The example of section 5 can be treated using Theorem IX.5 with $G = B(\rho)$ and ρ sufficiently large.

7. We shall consider here special cases of (IX.18) with $k = n = 1$, i.e. some second order scalar functional differential equations. The results we shall give remain valid if the right hand of the equation does depend on x', provided a suitable growth condition (such as Nagumo condition introduced in chapter V) with respect to x' is imposed on the equation at hand.

Let $f : R^3 \to R$, $(t,x,y) \to f(t,x,y)$ and $d : R^2 \to R$ be T-periodic in t and continuous and let us consider the scalar second order functional differential equation

$$x''(t) - f(t, x(t), x(t-d(t,x(t)))) = 0 \tag{IX.29}$$

Definition IX.2. The $\mathcal{C}^2$ T-periodic function α (resp. β) will be said to be a *lower* (resp. *upper*) *solution* of (IX.29) if, for each $t \in R$,

$$\alpha''(t) - f(t, \alpha(t), \alpha(t-d(t,\alpha(t)))) \leqslant 0$$

$$(\text{resp. } \beta''(t) - f(t, \beta(t), \beta(t-d(t,\beta(t)))) \leqslant 0) .$$

Theorem IX.6. *Let* f *be non increasing in* y *for each* (t,x) *and suppose there exists a lower solution* α *and an upper solution* β *of* (IX.29) *such that* $\alpha(t) \leqslant \beta(t)$, $t \in R$. *Then there exists a T-periodic solution* x *of* (IX.29) *such that*

$$\alpha(t) \leqslant x(t) \leqslant \beta(t) , \ t \in R.$$

The proof of this theorem will be accomplished by modifying equation (IX.29) in such a way that T-periodic solutions of the modified equations are also T-periodic solutions of (IX.29) and the modified equation can be easily treated by Corollary IV.1, an approach already used in Chapter V.

For $x \in R$ and fixed t, define $\overline{x}$ as follows

$$\overline{x} = \begin{cases} \beta(t) & \text{if } x > \beta(t) \\ x & \text{if } \alpha(t) \leqslant x \leqslant \beta(t) \\ \alpha(t) & \text{if } x < \alpha(t) \end{cases} \tag{IX.30}$$

and let

$$\overline{d}(t,x) = d(t,\overline{x}) , \tag{IX.31}$$

$$\overline{f}(t,x,y) = f(t,\overline{x},\overline{y}).$$

Further define

$$F(t,x,y) = \begin{cases} \overline{f}(t,x,y) + (x-\beta(t)) , & \text{if } x > \beta(t) \\ \overline{f}(t,x,y) & \text{if } \alpha(t) \leqslant x \leqslant \alpha(t) \\ \overline{f}(t,x,y) + (x-\alpha(t)) , & \text{if } x < \alpha(t) . \end{cases} \tag{IX.32}$$

Clearly F is T-periodic in t and continuous.

Let $\varepsilon > 0$ be given and let

$$A(t) = \alpha(t) - \varepsilon, \quad B(t) = \beta(t) + \varepsilon.$$

Lemma IX.2. *Let* x *be a T-periodic solution of*

$$(\overline{G}x)(t) \equiv x''(t) - F(t, x(t), x(t-\overline{d}(t,x(t)))) = 0. \qquad \text{(IX.33)}$$

Then, for each $t \in R$, $\alpha(t) \leq x(t) \leq \beta(t)$, *and hence* x *is a T-periodic solution of* (IX.29).

Proof. We have

$$(\overline{G}A)(t) = \alpha''(t) - F(t,\alpha(t)-\varepsilon,\alpha(t-\overline{d}(t,\alpha(t)-\varepsilon)) - \varepsilon)$$

$$= \alpha''(t) - \overline{f}(t,\alpha(t)-\varepsilon,\alpha(t-\overline{d}(t,\alpha(t)-\varepsilon) + \varepsilon$$

$$= \alpha''(t) - \overline{f}(t,\alpha(t)-\varepsilon,\alpha(t-d(t,\alpha(t)) - \varepsilon) + \varepsilon$$

$$= \alpha''(t) - f(t,\alpha(t),\alpha(t-d(t,\alpha(t))) + \varepsilon > 0$$

and, similarly, $(\overline{G}B)(t) < 0$, $t \in R$.

The proof will be complete if we show that $A(t) < x(t) < B(t)$, because $\varepsilon > 0$ is arbitrary. Let us show, say, that $x(t) < B(t)$, $t \in R$. If not, there will exist t such that $x(t) \geq B(t)$. Hence,

$$x''(t) - B''(t) > \overline{f}(t,x(t),x(t-\overline{d}(t,x(t))) + (x(t) - \beta(t)) - \overline{f}(t,\beta(t), B(t-\overline{d}(t,B(t))))- \varepsilon$$

$$= f(t,\beta(t),\overline{x}(t-d(t,\beta(t)))) + x(t) - \beta(t)$$

$$- f(t, \beta(t), \beta(t- d(t,\beta(t)))) - \varepsilon \geq x(t)-\beta(t)-\varepsilon \geq \beta(t)+\varepsilon-\beta(t)-\varepsilon=0.$$

Therefore, as long as $x(t) - B(t) \geq 0$, $x''(t) - B''(t) > 0$. As $x(t) - B(t)$ is T-periodic, this is impossible because if $x(\tau) - B(\tau)$ is a positive maximum of $x -$, then $x'(\tau)-B'(\tau)=0$, $x''(\tau) - B''(\tau) > 0$ and hence $x(t) - B(t) > x(\tau) - B(\tau)$ for $t > \tau$ and sufficiently close. Hence $x(t) < B(t)$ for each $t \in R$. The last assertion of the Lemma follows from the fact that f and F coincide if $\alpha(t) \leq x \leq \beta(t)$.

Lemma IX.3. *Let* $g : R \times R^n \to R^n$ *and* $d : R \times R^n \to R$ *be T-periodic in* t *continuous. Let* $\Omega \subset R^n$ *be a bounded open convex set containing* 0 *and assume that there exists an outer normal* n *to* Ω *such that, for every* $t \in R$, $y \in \overline{\Omega}$,

$$< n(x),f(t,x,y) > > 0 . \qquad \text{(IX.34)}$$

Then equation

$$x'' = g(t,x(t),x(t-d(t,x(t))))$$

has at least one T-periodic solution x *such that* $x(t) \in \Omega$, $t \in R$.

Proof. It is exactly the same than the one of Corollary V.22 in chapter V, and it is omitted.

Lemma IX.4. *There exists a constant* $M > 0$ *such that*

$$F(t, M, y) > 0 > F(t, -M, y) \ , \ t \ , \ y \in R \qquad (IX.35)$$

and equation (IX.33) *has at least one T-periodic solution* x *such that* $-M \leq x(t) \leq M \ , \ t \in R$.

Proof. (IX.35) follows at once from (IX.32) and the fact that $\overline{f}$ is bounded. The second part of the lemma follows from Lemma IX.3 applied to (IX.33) with $\Omega =]-M,M[\ , \ n(-M) = -1 \ , \ n(M) = +1$.

Proof of Theorem IX.6. By Lemmas IX.4 and IX.2, equation (IX.33) has at least one T-periodic solution x such that $\alpha(t) \leq x(t) \leq \beta(t) \ , \ t \in R$ which is therefore a T-periodic solution of (IX.29).

Theorem IX.7. *Let* $f : R^3 \to R$ *be a T-periodic in* t *and continuous and suppose that there exist* $\mathcal{C}^2$ *T-periodic functions* α, β *such that* $\alpha(t) \leq \beta(t) \ , \ t \in R$ *and such that for all* $y \in W = \{y \mid \alpha(s) \leq y \leq \beta(s), \ 0 \leq s \leq T\}$,

$$\alpha''(t) - f(t,\alpha(t),y) \geq 0 \geq \beta''(t) - f(t,\beta(t),y).$$

Then equation (IX.29) *has at least one T-periodic solution such that* $\alpha(t) \leq x(t) \leq \beta(t) \ , \ t \in R$.

Proof. It follows exactly the lines of the proof of Theorem IX.6 (the only slight modification is in Lemma IX.2) and is omitted.

Corollary IX.5. *Let there exist constants* α *and* $\beta, \ \alpha \leq \beta$ *such that either*

$$f(t,\alpha,\alpha) \leq 0 \leq f(t,\beta,\beta) \ ; \ t \in R$$

and $f(t,x,y)$ *is nonincreasing with respect to* y *or*

$$f(t,\alpha,y) \leq 0 \leq f(t,\beta,y) \ , \ t \in R \ , \alpha \leq y \leq \beta.$$

Then there exists a T-periodic solution x *of* (IX.29) *such that* $\alpha \leq x(t) \leq \beta \ , \ t \in R$.

Examples. The equation

$$x'' = x^3 - x(t-d(t, x(t)))) + e(t)$$

with e and d T-periodic in T and continuous satisfies the first condition of Corollary IX.5 with $\beta > 1$ and

$$\beta(\beta^2-1) - \max_{t\in[0,T]} |e(t)| > 0 \ \text{ and } \alpha < -1 \text{ and } \alpha(\alpha^2-1) + \max_{t\in[0,T]} |e(t)| < 0$$

The equation

$$x'' = x + \sin x(t - d(t,x)) + e(t)$$

satisfies the second condition of Corollary IX.5.

8. Let us now come back to equation (IX.18) and, for simplicity, suppose that $n = 1$ (scalar case), and that (IX.18) has the form

$$x^{(k+1)} + a_1 x^{(k)} + \dots + a_k x' = g(t,x(t-r)) \qquad \text{(IX.36)}$$

where $y : R \times R \to R$ is T-periodic in t and continuous and where the $a_i (i=1,\dots,k)$ are constants. We shall further assume that constants $\mu',\nu' \geq 0$ and $\delta \in [0,1[$

$$|g(t,y)| \leq \mu'|y|^{\delta} + \nu' \qquad \text{(IX.37)}$$

for all $t \in R$ and $y \in R$.

Theorem IX.8. *Suppose that for each* $t \in R$,

$$\lim_{y \to +\infty} \frac{g(t,y)}{|y|^{\delta}} = g_+(t) \qquad \text{(IX.38)}$$

and

$$\lim_{y \to -\infty} \frac{g(t,y)}{|y|^{\delta}} = g_-(t) \qquad \text{(IX.39)}$$

exist and are such that

$$\int_0^T g_- < 0 < \int_0^T g_+ \qquad \text{(IX.40)}$$

or

$$\int_0^T g_- > 0 > \int_0^T g_+ \qquad \text{(IX.41)}$$

Then equation (IX.36) *has at least one* T*-periodic solution.*
If moreover $\delta = 0$ *and for* $t \in R$, $y \in R$,

$$g_-(t) \leq g(t,y) \leq g_+(t) \qquad \text{(IX.42)}$$

or

$$g_-(t) \geq g(t,y) \geq g_+(t) \quad) \qquad \text{(IX.43)}$$

with the strict inequalities holding for a set of values of s *of positive measure, then* (IX.40) (*or* (IX.41)) *is also necessary for the existence of a* T*-periodic solution for* (IX.36).

Proof. We shall apply Theorem VII.4 to the operator equation in P_T corresponding to (IX.36). Clearly the regularity assumptions for L and N as well a condition (a) of Theorem VII.4 are satisfied.

If condition (b) does not hold, then there exists a bounded $V \subset \operatorname{Im} L$, a sequence (t_n) with $t_n > 0$ $(n \in \mathbb{N})$ and $t_n \to \infty$ if $n \to \infty$, a sequence (z_n) with $z_n \in V$ and a sequence (w_n) with $w_n \in \ker L \cap \partial B(1)$ such that

$$\int_0^T g(\tau, t_n w_n + t_n^\delta z_n(\tau - r) d\tau = 0 .$$

As $\ker L \cap \partial B(1) = \{-1, 1\}$ we can assume by taking a subsequence if necessary that $w_n = +1$ or $w_n = -1$ for all n. Consider for definiteness the case where $w_n = 1$.

Thus

$$\int_0^T t_n^{-\delta} g(\tau, t_n + t_n^\delta z_n(\tau - r)) d\tau = 0 \qquad \text{(IX.44)}$$

We have, because of $z_n \in V$ for all n,

$$1 + t_n^{\delta - 1} z_n(\tau - r) \to 1$$

if $n \to \infty$ uniformly in τ, and hence, for sufficiently large n,

$$\frac{g(\tau, t_n + t_n^\delta z_n(\tau - r))}{t_n^\delta} =$$

$$\frac{g(\tau, t_n + t_n^\delta z_n(\tau - r))}{|t_n + t_n^\delta z_n(\tau - r)|^\delta} \, |1 + t_n^{\delta - 1} z_n(\tau - r)| \to g_+(\tau)$$

if $n \to \infty$. By (IX.38) (IX.39) and (IX.37), g_+ (and g_-) are in $L^1(0, 2\pi)$ and it follows from (IX.44) and Lebesgue convergence theorem that

$$\int_0^T g_+(\tau) d\tau = 0 ,$$

a contradiction with (IX.40). Thus condition (b) of Theorem (VII.4) holds. Now if $a \in \mathbb{R}$, $a \neq 0$, it follows from (IX.38) and (IX.39) and Lebesgue convergence theorem that

$$\frac{1}{T} \int_0^T \frac{g(t, a)}{|a|^\delta} dt \to \frac{1}{T} \int_0^T g_\pm$$

if $a \to \pm \infty$ and hence if say (IX.40) holds, there exists $R > 0$ such that

$$\frac{1}{T}\int_0^T g(t, -R)dt < 0 < \frac{1}{T}\int_0^T g(t,R)dt$$

which implies that condition (c) of Theorem (VII.4) holds and achieves the proof of sufficiency. Now if (IX.36) has a T-periodic solution x then necessarily

$$\int_0^T g(t, x(t-r))dt = 0$$

and hence by, say, (IX.42) ,

$$\int_0^T g_-(t)dt < 0 < \int_0^T g_+(t)dt$$

the strict inequalities following from the fact that inequalities hold in (IX.42) on a set of positive measure in (0,T).

9. Bibliographical notes about Chapter IX.

Theorem IX.1 is due to Fabry and Franchetti (*J. Differential Equations*, to appear) (see also the *Seminario dell' Istituto di Matematica Applicata, Firenze*, december 1973 and january 1974), as well as Corollaries IX.1, IX.2 and Theorem IX.2 which generalizes to the case $\alpha \neq 0$ a former result of Lazer and Leach (*Ann. Mat. Pura Appl.* (4) 82 (1969) 49-68). Theorem IX.3 extends in some directions a recent result of Reissig (*Atti Accad. Naz. Lincei, Rend. Cl. Sci. Mat. Fis. Nat.*, to appear) where on the other side some of the assumptions are given in a sharper way by taking in account the particular structure of the equation. See also for a related problem Sedsiwy (*Proc. Amer. Math. Soc.*, to appear). For other applications of the results of Chapter VII, and in particular of Corollary VII.2 to equations of type (IX.13) and corresponding retarded functional differential equations, see Mawhin (*J. Math. Anal. Appl.* 45 (1974) 588 - 603) which generalizes in particular results of Ezeilo (*Proc. Cambridge Phil. Soc.* 56 (1960) 381-389), Fennell (*J. Math. Anal. Appl.* 39(1972) 198 - 201), Lazer (*J. Math. Anal. Appl.* 21(1968) 421 - 425 , Reissig (*Ann. Mat. Pura Appl.* (4) 85 (1970), 235 - 240; 87 (1970) 111 - 124; *Atti. Accad. Naz. Lincei, Rend. Cl. Sci. Fis. Mat. Nat.* (8) 48 (1970) 484 - 486), Sedsiwy (*Ann. Polon. Math.* 17 (1965) 147 - 154; *Zeszyty Nauk. Univ. Jagiell.* N°131 (1966) 69 - 80; *Atti. Accad. Naz. Lincei, Rend. Cl. Sci. Fis. Mat. Nat.* (8) 47 (1969) 472 - 475) and Villari (*Ann. Mat. Pura Appl.* (4) 73 (1966) 103 - 110).

The proof of Lemma IX.1 is inspired by J.K. Hale ("Functional differential equations", Springer, New York, 1971, ch III), a very good general reference on the subject of functional differential equations. The concept of guiding function in section 5 is an extension of the one given by Kranosel'skii ("The operator of translation along trajectories of ordinary differential equations", Amer. Math. Soc., Providence, 1968) for ordinary and delay-differential equations. The fact that only t such that $|V(x(t))| \geq |V(x(s))|$, $s \in [0,T]$ have to be considered in this definition was suggested to us by O. Lopes (Providence, 1973, personal communication). Theorem IX.4 extends Krasnosel'skii results with a much more simple proof (Krasnosel'skii used the translation operator). The case of equation (IX.21) was given in Mawhin (*J. Differential Equations* 10(1971) 240 - 261) where other existence theorems for periodic solutions of functional differential equations are also proved. See also Krasnosel'skii and Lifchits (*Automatike i Telemekhanike* 9 (1973) 12 - 15 for a recent account of guiding functions.

Theorem IX.5 generalizes, with a much simpler proof, a result of G. B. Gustafson and K. Schmitt (*Proc. Amer. Math. Soc.* 42 (1974) 161 - 166) given in Corollary IX.4. Theorem IX.6 and IX.7 are due to K. Schmitt ("Equations différentielles et fonctionnelles non linéaires", Hermann, Paris, 1973 65 -78) but the proof given here is simplified by the direct use of Lemma IX.3, first proved by G. B. Gustafson and K. Schmitt (*J. Differential Equations* 13(1973) 567 - 587) where also other existence theorems for functional differential equations, of local nature and which can be got using Theorem IX. 2, can be found.

Theorem IX.8 generalizes a result of Fučik (*Czechoslovak Math. J.* 24(99) 1974, 467 - 495) who considers the case where $k = \delta = 0$ and gives also other applications of a special case of Theorem VII.4.

For other results using coincidence degree theory for periodic solutions of functional differential equations, see Cronin (*Bol. Un. Mat. Ital.* (4)6(1972) 45 - 54; *J. Differential Equations* 14(1973) 581 - 596), Hale and Mawhin (*J. Differential Equations*, 15 (1974), 295 - 307 where the case of neutral equations is treated and Knolle (to appear) who considers a population problem. For applications of some of the results of Chapter VII to boundary value problems for ordinary differential equations see S.Fučik (Boundary value problems with jumping nonlinearities, to appear

In the search of periodic solutions for some differential equations containing quasibounded nonlinearities together with other ones, the simultaneous use of L^2-estimates and degree techniques leads to interesting results. See Mawhin (*J. Math. Anal. Appl.* 40(1972) 20 - 29), Fučik and Mawhin (*Časopis pro pěst. matemat.* 100 (1975) 276 - 283), Reissig (*Ann. Mat. Pure Appl.* (4) 104 (1975) 269 - 281; *Atti. Acc. Naz. Lincei, Rend. Cl. Sci. fis. mat. natur.* (8) 61 (1974) 297 - 302; *Abh. Math. Sem. Univ. Hamburg*, to appear).

. COINCIDENCE INDEX, MULTIPLICITY AND BIFURCATION THEORY

. Let X,Z be normed real vector spaces, $a \in X$ and Ω a bounded open neighbour-
ood of a. Let $L : \text{dom } L \subset X \to Z$ be a (not necessarily continuous) linear mapping nd $N : \Omega \to Z$ a (not necessarily linear) continuous mapping such that $N(\Omega)$ is ounded, which satisfy the following assumptions.

(H_1)L *is a Fredholm mapping of index zero*, i.e. dim ker L = codim Im L $< \infty$ nd Im L is closed in Z.

(H_2)N *is L-compact on* $\overline{\Omega}$.

(H_3) *The point* a *is an isolated zero of* L-N.

It follows from assumption (H_3) that there exists an $\varepsilon_0 > 0$ such that ne closure $\overline{B}_{\varepsilon_0}(a)$ of the open hall $B_{\varepsilon_0}(a)$ of center a and radius ε_0 is contai-ed in $\overline{\Omega}$ and such that

$$\{(L - N)^{-1}(0)\} \cap \overline{B}_{\varepsilon_0}(a) = \{0\}.$$

ence, for every $\varepsilon \in]0,\varepsilon_0[$,

$$0 \notin (L - N)(\text{dom } L \cap \partial B_\varepsilon(a))$$

nd it follows then from chapter III that the coincidence degree $d[(L,N),B_\varepsilon(a)]$ of and N in $B_\varepsilon(a)$ will be defined and will not depend upon ε. The invariance of $[(L,N),B_\varepsilon(a)]$ with respect to ε justifies the following

efinition X.1. Assumptions $(H_1\text{-}H_2\text{-}H_3)$ being supposed satisfied, the *coincidence ndex* $i[(L,N),a]$ *of* L *and* N *at point* a is the integer $d[(L,N), B_\varepsilon(a)]$ for any $\in]0,\varepsilon_0[$.

It follows at once from Definition X.1 that

$$i[(L,N),a] = i_{LS}[I - M_\Lambda, a]$$

here the right hand member denotes the Leray-Schauder index of the fixed point a M_Λ.

An interesting property of the coincidence index showing its interest in the omputation of coincidence degree is the following

roposition X.1. *Let* $L : \text{dom } L \subset X \to Z$, $N : \text{cl}\Sigma \to Z$ *be mappings satisfying conditions* $(H_1\text{-}H_2)$ *above with* Ω *replaced by the open bounded subset* Σ *of* X. *If*

$$0 \notin (L - N)(\text{dom } L \cap \partial\Sigma)$$

d if $(L-N)^{-1}(0)$ *is a finite set* $\{a_1,\ldots,a_m\}$, *then*

$$d[(L,N), \Sigma] = \sum_{j=1}^{m} i[(L,N),a_j] \qquad (X.1)$$

Proof. Formula (X.1) is an easy consequence of Definition X.1 and of the additivity and excision properties of coincidence degree proved in chapter III.

2. We shall be interested in this section in founding various conditions under which the coincidence index of L and N at some point, say the origin, can be estimated. In the whole section, $\Omega \subset X$ will be an open bounded neighbourhood of the origin.

Theorem X.1. *Let* L *and* N *satisfy assumptions* $(H_1 - H_2)$ *above with* $N(0) = 0$ *and let* L-N *be one-to-one on* Ω . *Then* $i[(L,N),0]$ *exists and*

$$|i[(L,N),0]| = 1. \qquad (X.2)$$

Proof. Let $B_\varepsilon(0)$ be such that $\overline{B}_\varepsilon(0) \subset \Omega$. It is clear that $(L-N)^{-1}(0) = \{0\}$ and that $i[(L,N),0]$ exists. By Proposition II.2 $I-M_\Lambda$ is one-to-one on $\overline{B}_\varepsilon(0)$ and hence, by a well-known theorem of Leray

$$d_{LS}[I-M_\Lambda, B_\varepsilon(0), 0] = \pm 1$$

which implies (X.2).

A more precise result will now be given, which needs more assumptions upon N, and which proof makes use of the following

Lemma X.1. *Suppose that* L *satisfies assumptions* (H_1) *and that* $N = A + B$ *with* $A : X \to Z$ *linear, L-compact on bounded sets and* $\ker(L-A) = \{0\}$, *and with* $B : \Omega \to Z$ *is L-compact on* Ω . *Then* $(L-A) : \operatorname{dom} L \to Z$ *is onto,* $(L-A)^{-1}B$ *is compact on* Ω *and, for every* $x \in \Omega$,

$$(I-M_\Lambda)x = (I-M_\Lambda^*)[I-(L-A)^{-1}B]x \qquad (X.3)$$

where

$$M_\Lambda^* = P + \Lambda\Pi A + K_{P,Q}A .$$

Proof. From Proposition III.2 we get at once that $\ker[I-M_\Lambda^*] = \{0\}$ and hence, using Riesz theory, $I-M_\Lambda^*$ will be a linear homeomorphism of X, and hence of dom L because of the form of $I - M_\Lambda^*$. Moreover using Lemma III.4,

$$(I-M_\Lambda^*)x = [\Lambda\Pi + K_{P,Q}](L-A)x$$

for each $x \in \operatorname{dom} L$ and hence, $\Lambda\Pi + K_{P,Q}$ being an algebraic isomorphism between Z and dom L, we have

$$(L-A)x = [\Lambda\Pi + K_{P,Q}]^{-1}(I-M_\Lambda^*)x$$

for each $x \in \operatorname{dom} L$. Therefore L-A is onto and, for each $z \in Z$,

$$(L-A)^{-1}z = (I-M_\Lambda^*)^{-1}[\Lambda\Pi + K_{P,Q}]z \qquad (X.4)$$

which clearly shows moreover that $(L-A)^{-1}B$ is compact on Ω .

Relations (X.3) then follows from (X.4) by direct computation.

If B is a mapping from some neighbourhood of 0 in X into Z and L is like above, we will write that

$$Bx = o_L(\|x\|)$$

if $\Pi Bx = o(\|x\|)$ and $K_{P,Q}Bx = o(\|x\|)$ with the usual meaning for $o(\|x\|)$.

It is easily checked that this definition does not depend upon P and Q and reduces to

$$Bx = o(\|x\|)$$

if $X = Z$ and $L = I$. It will be in particular satisfied if $Bx = o(\|x\|)$ and $K_{P,Q}$ is continuous

Theorem X.2. *Let us suppose that* L *and* N *satisfy assumptions* (H_1) *and* (H_2) *above, that*

$$N = A + B$$

with $A : X \to Z$ *linear and* $B : \overline{\Omega} \to Z$ *such that* $\ker (L - A) = \{0\}$ *and*

$$Bx = o_L(\|x\|). \tag{X.5}$$

Then, 0 *is an isolated zero of* $L - N$ *and*

$$i [(L,N),0] = i [(L,A),0].$$

Proof. First using Lemma X.1 we obtain

$$(I-M_\Lambda)x = (I-M_\Lambda^*)[I - (L-A)^{-1}B]x \tag{X.6}$$

for every $x \in \overline{\Omega}$, with $(L-A)^{-1}B$ compact in $\overline{\Omega}$. Now, by (X.5) and (X.4) there exists $\varepsilon_1 > 0$ such that $\overline{B}_{\varepsilon_1}(0) \subset \overline{\Omega}$ and

$$\|(L-A)^{-1} Bx\| \leq (\tfrac{1}{2}) \|x\|$$

for every $x \in \overline{B}_{\varepsilon_1}(0)$. Hence, for every $\varepsilon \in]0,\varepsilon_1]$ and every $(x,\lambda) \in \overline{B}(0) \times [0,1]$, we have

$$\|x - \lambda (L-A)^{-1} Bx\| \geq (\tfrac{1}{2}) \|x\|$$

which first shows that 0 is an isolated zero of L-N and then, using the invariance of degree with respect to homotopy, that

$$d_{LS}[I-\lambda(L-A)^{-1} B, B_\varepsilon(0), 0] = 1$$

for every $\lambda \in [0,1]$. Now, using (X.6), (X.7) and Leray's product theorem we obtain

$$i [(L,N),0] = d_{LS}[I-M , B(0),0] = d_{LS}[I-M , B (0),0] = i[(L,A),0]$$

which achieves the proof.

We shall now be interested in computing $i[(L,N),0]$ when the linear part of L-N has a nontrivial kernel. Typical for this situation is the case where $\ker L \neq \{0\}$ and N contains no linear terms.

We shall suppose therefore that, n and k being non negative integers, one has (H_4) $N = \sum_{i=0}^{k} C_{n+i} + R$, *where, for each* i, C_{n+i} *is a continuous, homogenous mapping of order* n+i *and* $R : \Omega \to Z$ *a mapping such that*

$$Rx = o_L(\|x\|^{n+k}).$$

We prove now the main theorem.

<u>Theorem X.3</u>. : *Let* (L,N) *satisfy assumptions* $(H_1),(H_2),(H_4)$ *and be such that :*

(i) $i \neq k$, $C_{n+i}(\ker L) \subset \operatorname{Im} L$.

(ii) $\Lambda\Pi C_{n+k}x \neq 0$; $\forall x \in \ker L$, $\|x\| = 1$.

(iii) *there exists* $\alpha > 0$ *such that,* $\forall i \neq k$, $\forall x, y \in \bar{\Omega}$,

$$\|\Lambda\Pi(C_{n+i}x - C_{n+i}y)\| + \|K_{P,Q}(C_{n+i}x - C_{n+i}y)\| < \alpha \|x - y\| \max(\|x\|^{n+i-1}, \|y\|^{n+i-1}).$$

(iv) $C_{n+k}(B_1(0))$ *is bounded in* Z .

(v) $n - 1 > k \geqslant 0$.

Then, $i[(L,N),0]$ *is well defined and*

$$i[(L,N),0] = i_B[-\widetilde{\Lambda\Pi C}_{n+k},0]$$

where the subscript B *denotes the Brouwer index and where* $\widetilde{\Lambda\Pi C}_{n+k}$ *is the restriction* $\Lambda\Pi C_{n+k}|\ker L$ *of* $\Lambda\Pi C_{n+k} : X \to \ker L$ *to* $\ker L$.

<u>Proof</u> : Consider, for each $\lambda \in [0,1]$, the operator $M_\lambda : \bar{\Omega} \to X$ defined by

$$M_\lambda = P + \Lambda\Pi C_{n+k} + \lambda\Lambda\Pi\left[\sum_{i=0}^{k-1} C_{n+i} + R\right] + \lambda K_{P,Q}N$$

It is clear that $M : [0,1] \times \Omega \to X$; $(\lambda,x) \to M_\lambda x$ is compact on $[0,1] \times \bar{\Omega}$ and such that

$$M(1,.)=M_\Lambda \quad\text{and}\quad M(0,.)=P + \Lambda\Pi C_{n+k} .$$

Let x_λ be a possible fixed point of M_λ such that $\|x_\lambda\|=\varepsilon > 0$

x_λ must verify

$$(I-P)x_\lambda = \lambda K_{P,Q} Nx_\lambda$$

and by (iii), (iv) and H_4, there exist an $\varepsilon_1>0$, $p>0$,

such that, for every $\varepsilon\leqslant\varepsilon_1$, we have

$$\sup_{\|x\|=\varepsilon} \|K_{P,Q} Nx\| \leqslant \lambda p\varepsilon^n$$

and then x_λ must be such that $\|(I-P)x_\lambda\| \leqslant \lambda p\varepsilon^n$. But if $\varepsilon_2 > 0$ is such that

$2p\varepsilon_2^{n-1} < 1$, then for every $\varepsilon \leqslant \min(\varepsilon_1,\varepsilon_2)$, we have also

$$\|Px_\lambda\| \geqslant \|x_\lambda\| - \|(I-P)x_\lambda\| \geqslant \frac{\varepsilon}{2} \geqslant p\varepsilon^n \geqslant \|(I-P)x_\lambda\|$$

and

$$\|Px_\lambda\| \leqslant \|x_\lambda\| + \|(I-P)x_\lambda\| \leqslant \frac{3\varepsilon}{2} .$$

But x_λ must also verify the equation

$$\Lambda\Pi C_{n+k}\, x_\lambda + \lambda\Lambda\Pi[\sum_{i=0}^{k-1} C_{n+i} + R]\, x_\lambda = 0$$

and, since for every $i \neq k$, $\Lambda\Pi C_{n+i}\, Px_\lambda = 0$

and $\|x_\lambda\| = \varepsilon \neq 0$,

$$\|x_\lambda\|^{1-n-k}\, \Lambda\Pi C_{n+k} x_\lambda = -\lambda\|x_\lambda\|^{1-n-k}\, \Lambda\Pi[\sum_{i=0}^{k-1} (C_{n+1}x_\lambda - C_{n+i}Px_\lambda) + Rx_\lambda]$$

which is equivalent to

$$(I-P)x_\lambda - \|x_\lambda\|^{1-n-k}\, \Lambda\Pi C_{n+k}x_\lambda = (I-P)x_\lambda$$

$$+ \lambda\, \|x_\lambda\|^{1-n-k}\, \Lambda\Pi[\sum_{i=0}^{k-1} (C_{n+i}x_\lambda - C_{n+i}Px_\lambda) + Rx_\lambda]$$

If we note that

$$\|(I-P)x_\lambda - \|x_\lambda\|^{1-n-k}\Lambda\Pi C_{n+k}x_\lambda\| = \|x_\lambda\|\, \|(I-P)y_\lambda - \Lambda\Pi C_{n+k}y_\lambda\|$$

where $y_\lambda = x_\lambda/\|x_\lambda\|$ and that by (ii), $(I-P)y_\lambda - \Lambda\Pi C_{n+k}y_\lambda$

must always be different from zero, we have by a classical argument

$\inf\limits_{\|y\|=1} \|(I-P)y - \Lambda\Pi C_{n+k}y\| \geqslant \gamma > 0$. So we conclude

$$\varepsilon\gamma \leqslant \|(I-P)x_\lambda - \|x_\lambda\|^{1-n-k}\, \Lambda\Pi C_{n+k}x_\lambda\| \leqslant \|(I-P)x_\lambda\| + \lambda\,\alpha \sum_{i=0}^{k-1} (\tfrac{3}{2})^{n+i-1}\, \varepsilon^{i-k}\, \|(I-P)x_\lambda\| +$$

$$+ \lambda\, \varepsilon^{1-n-k}\, q(\varepsilon)\, \|x_\lambda\|^{n+k}$$

where $q(\varepsilon) \to 0$ si $\varepsilon \to 0$. Hence if $0 < \varepsilon \leqslant \min(\varepsilon_1,\varepsilon_2)$, $\gamma \leqslant p\varepsilon^{n+1} +$
$\lambda\,\alpha p \sum\limits_{i=0}^{h-1} (\frac{3}{2})^{n+i-1}\, \varepsilon^{n-k-1+i} + \lambda q(\varepsilon)$.

But there exists $0 < \varepsilon_3 \leqslant \min(\varepsilon_1,\varepsilon_2)$ such that for $0 < \varepsilon \leqslant \varepsilon_3$ and $\lambda \in [0,1]$ one has

$$p\varepsilon^{n+1} + \lambda\alpha p \sum_{i=1}^{k-1} (\tfrac{3}{2})^{n+i-1}\, \varepsilon^{n-k-1+i} + \lambda q(\varepsilon) < \gamma$$

and therefore M_λ for each $\lambda \in [0,1]$ cannot have a fixed point x_λ such that

$\|x_\lambda\| = \varepsilon$, $0 < \varepsilon \leq \varepsilon_3$. Thus 0 is an isolated zero of $(L - N)$, and

$$i[(L,N),0] = i_{LS}[I-M_1,0] = i_{LS}[I-M_0,0] = i_{LS}[I-P-\Lambda\Pi C_{n+k},0] =$$

$$i_B[-\Lambda\Pi C_{n+k},0]$$

which completes the proof.

3. Let X, Z be here two normed vector spaces on the complex field, $L : \text{dom } L \subset X \to Z$ and $A : X \to Z$ be linear mappings satisfying (H_1) for L and :

(H_5) A *is L-compact.*

Definition X.2. If (L,A) satisfies the assumptions (H_1) and (H_5) above, $\mu \in C$ is said to be a *characteristic value for* (L,A) if

$$\ker (L - \mu A) \neq \{0\}$$

and a *regular covalue for* (L,A) is the mapping

$$L - \mu A : \text{dom } L \subset X \to Z$$

has a continuous inverse.

For each $\mu \in C$, the linear operator

$$M(\mu) = P + \mu(\Lambda\Pi + K_{P,Q})A \tag{X.8}$$

is compact and, as shown in Proposition III.2 and Lemma III.4,

$$\ker(L - \mu A) = \ker [I - M(\mu)], \tag{X.9}$$

and, for $x \in \text{dom } L$,

$$[I - M(\mu)]x = (\Lambda\Pi + K_{P,Q})(L - \mu A)x, \tag{X.10}$$

$\Lambda\Pi + K_{P,Q}$ being a linear isomorphism of Z onto dom L.

Also it follows from the compactness of $M(\mu)$ and of (X.9) that $I - M(\mu)$ has a continuous inverse when μ is not a characteristic value for (L,A) and, because of the relation deduced from (X.4)

$$(L - \mu A)^{-1} = [I - M(\mu)]^{-1}(\Lambda\Pi + K_{P,Q}), \tag{X.11}$$

$(L - \mu A)^{-1}$ is continuous together with $K_{P,Q}$.

In order to be able to define a multiplicity for the characteristic values for (L,A) we shall make the following supplementary assumption

(H_6) *The set of characteristic values for* (L,A) *is not* C .

An useful sufficient condition insuring that (H_6) holds will be given later.

Definition X.3. If the pair (L,A) satisfies assumptions (H_1), (H_5), (H_6) a *spectral operator for* (L,A) will be any operator $A_0 : X \to X$ of the form

$$A_0 = (L - \mu_0 A)^{-1} A$$

where μ_0 is not a characteristic value for (L,A).

The interest of this concept is that it allows the reduction of the study of the spectral properties of (L,A) to that of the compact mapping A_0, as follows from

Lemma X.2. *If assumptions* (H_1), (H_5), (H_6) *hold and if* A_0 *is the spectral operator for* (L,A) *associated to* μ_0, *then*

(i) A_0 *is compact;*

(ii) *the set of the characteristic values for* (L,A) *is the translated by* μ_0 *of the set of the characteristic values, in the classical sense, of* A_0.

Proof. The results follow at once from the relation deduced from (X.3)

$$I - M(\mu) = [I - M(\mu_0)]\,[I - (\mu - \mu_0)A_0] \tag{X.12}$$

and from assumption (H_5) since, by (X.11)

$$A_0 = [I - M(\mu_0)]^{-1}(\Lambda\Pi + K_{P,Q})A .$$

If we suppose now that $K_{P,Q} : Z \to X$ is continuous, an assumption independent upon the choice of P and Q, we obtain more information about the spectral properties of (L,A), which follows at once from (X.11) and the properties of linear compact mappings.

Lemma X.3. *If* (L,A) *satisfies assumptions* (H_1), (H_5) *and if* L *is such that* $K_{P,Q}$ *is continuous, then each* $\mu \in C$ *is either a characteristic value or a regular covalue for* (L,A), *and the set of characteristic values is either* C *or a set at most countable with a possible accumulation point at infinity.*

In order to use the concept of spectral operator for defining a reasonable concept of multiplicity for the characteristic values of (L,A) we need the following

Lemma X.4. *If assumptions* (H_1), (H_5), (H_6) *hold and if* μ *is a characteristic value for* (L,A) *then, for any* μ_0 *which is not a characteristic value for* (L,A) *the classical multiplicity of* $\mu - \mu_0$ *as a characteristic value of the spectral operator* A_0 *is defined and does not depend upon* μ_0.

Proof. It follows from Lemma X.2 that if μ (resp. μ_1 and μ_2) is (resp. are not) characteristic value(s) for (L,A), then the operators

$$A_i = (L - \mu_i A)^{-1} A \quad (i = 1,2)$$

are compact and respectively admit $\mu - \mu_i$ $(i = 1,2)$ as characteristic values with multiplicities

$$\beta_i(\mu) = \dim \ker [I - (\mu - \mu_i)A_i]^{n_i(\mu)} \quad (i = 1,2)$$

where $n_i(\mu)$ is the smallest integer such that

$$\ker [I - (\mu - \mu_i)A_i]^{n+1} = \ker [I - (\mu - \mu_i)A_i]^n \quad (i = 1,2).$$

Now relation (X.12) implies that

$$I - M(\mu) = [I - M(\mu_i)] [I - (\mu - \mu_i)A_i] \quad (i = 1,2)$$

and

$$I - M(\mu_1) = [I - M(\mu_2)] [I - (\mu_1 - \mu_2)A_2].$$

$I - M(\mu_i)$ $(i = 1,2)$ being invertible, as well as $I - (\mu_1 - \mu_2)A_2$

we have

$$I - (\mu - \mu_1)A_1 = [I - M(\mu_1)]^{-1} [I - M(\mu)] =$$
$$= [I - (\mu_1 - \mu_2)A_2]^{-1} [I - (\mu - \mu_2)A_2].$$

Then, using the relation

$$[I - (\mu_1 - \mu_2)A_2]^{-1} [I - (\mu - \mu_2)A_2] = [I - (\mu - \mu_2)A_2] [I - (\mu_1 - \mu_2)A_2]^{-1},$$

we obtain, for each positive integer n,

$$[I - (\mu - \mu_1)A_1]^n = \{[I - (\mu_1 - \mu_2)A_2]^{-1} [I - (\mu - \mu_2)A_2]\}^n =$$
$$= [I - (\mu_1 - \mu_2)A_2]^{-n} [I - (\mu - \mu_2)A_2]^n.$$

Hence, for each integer $n > 0$,

$$\ker [I - (\mu - \mu_1)A_1]^n = \ker [I - (\mu - \mu_2)A_2]^n,$$

and the proof is complete.

Lemma X.4 justifies the following

Definition X.4. If assumptions (H_1), (H_5), (H_6) hold for the pair (L,A) and if $A_0 = (L - \mu_0 A)^{-1}A$ is any spectral operator for (L,A), the *multiplicity* $\beta(\mu)$ *of the characteristic value* μ *for* (L,A) is the integer

$$\beta(\mu) = \dim \ker [I - (\mu - \mu_0)A_0]^{n(\mu)}.$$

where $n(\mu)$ is the smallest nonnegative integer such that

$$\ker [I - (\mu - \mu_0)A_0]^n = \ker [I - (\mu - \mu_0)A_0]^{n+1}.$$

This definition agrees with the classical one when X = Z and L = I because, for the pair (I,A), 0 is not a characteristic value and hence A_0 can be chosen to be A.

An interesting situation in which the multiplicity can be more easily computed is given by the following

<u>Proposition X.2</u>. *Let* (L,A) *satisfy assumptions* (H_1), (H_5), (H_6) *and let* μ *be a characteristic value for* (L,A) . *Then*

$$\beta(\mu) = \dim \ker (L - \mu A)$$

if and only if

$$A\,[\ker(L - \mu A)] \cap \mathrm{Im}\,(L - \mu A) = \{0\} .$$

<u>Proof</u>. It follows from the definition that the multiplicity $\beta(\mu)$ of the characteristic value μ for (L,A) will be equal to the dimension of $\ker (L - \mu A)$ if and only if

$$\ker [I - (\mu - \mu_0)A_0] = \ker [I - (\mu - \mu_0)A_0]^2 .$$

i.e. using Riesz theory if and only if

$$\ker [I - (\mu - \mu_0)A_0] \cap \mathrm{Im}\,[I - (\mu - \mu_0)A_0] = \{0\}.$$

But the following sequence of equivalences is easily verified

$$\ker [I - (\mu - \mu_0)A_0] \cap \mathrm{Im}\,[I - (\mu - \mu_0)A_0] = \{0\} \Leftrightarrow$$

$$\Leftrightarrow \ker (L - \mu A) \cap \mathrm{Im}\,(L - \mu_0 A)^{-1}(L - \mu A) = \{0\} \Leftrightarrow$$

$$\Leftrightarrow (L - \mu_0 A) \ker (L - \mu A) \cap \mathrm{Im}\,(L - \mu A) = \{0\} \Leftrightarrow$$

$$\Leftrightarrow (\mu - \mu_0)A\,[\ker(L - \mu A)] \cap \mathrm{Im}(L - \mu A) = \{0\} ,$$

and hence the proof is complete.

Let us note that assumption (H_6) implies that

$$\ker L \cap \ker A = \{0\} ,$$

which in turn gives

$$\ker(L - \mu A) \cap \ker A = \{0\}$$

for each $\mu \in R$. Then the condition of Proposition X.2 is equivalent to the condition

$$\forall\, x \in \ker (L - \mu A) \{0\},\ Ax \notin \mathrm{Im}(L - \mu A).$$

The definition of multiplicity given above is rather implicit, in that it requires the introduction of the operator A_0 . We shall see now that, if we replace (H_6) by a stronger condition, we can relate closely the multiplicity to a particular operator $M(\mu)$. This stronger condition is the following *transversality condition*

(H_6') $\forall x \in \ker L \setminus \{0\}$, $Ax \notin \operatorname{Im} L$.

That condition (H_6') implies (H_6) will be shown in the sequel. First we need some technical lemmas, in which (L,A) is supposed to satisfy the assumptions (H_1), (H_5), (H_6').

<u>Proposition X.3</u>. *Under assumptions above, there exists an unique (continuous) projector* $Q_A : Z \to Z$ *such that*

$$\operatorname{Im} Q_A = A(\ker L) \ , \ \ker Q_A = \operatorname{Im} L \ .$$

<u>Proof</u>. By (H_6'), $A|\ker L$ is bijective and hence

$$\dim A(\ker L) = \dim \ker L = \dim \operatorname{coker} L \ .$$

Moreover, it follows from the relation

$$A(\ker L) \cap \operatorname{Im} L = \{0\}$$

that A(ker L) is then an algebraic, and hence topological, supplement of Im L in Z. The existence and unicity of Q_A follows at once.

<u>Proposition X.4</u>. *Under assumptions above, the mapping*

$$P_A : X \to X, \quad x \mapsto (\Pi A|\ker L)^{-1} \Pi Ax$$

is a (continuous) projector on X *such that* $\operatorname{Im} P_A = \ker L$ *and, for each* $x \in X$,

$$\Pi A(I - P_A)x = 0 \ . \tag{X.13}$$

<u>Proof</u>. It is clear that the mapping $\Pi A|\ker L : \ker L \to \operatorname{coker} L$ is an algebraic isomorphism and hence $(\Pi A|\ker L)^{-1}$ is well defined. Moreover,

$$\begin{aligned} P_A^2 x &= (\Pi A \ker L)^{-1} | \Pi A(\Pi A|\ker L)^{-1}\Pi Ax \\ &= (\Pi A|\ker L)^{-1}(\Pi A|\ker L)\ (\Pi A|\ker L)^{-1}\Pi Ax = P_A x \end{aligned}$$

and

$$\begin{aligned} \Pi A(I - P_A)x &= \Pi Ax - \Pi A(\Pi A|\ker L)^{-1}\Pi Ax \\ &= \Pi Ax - (\Pi A|\ker L)\ (\Pi A|\ker L)^{-1}\Pi Ax = 0 \ , \end{aligned}$$

which achieves the proof.

With those projectors we shall introduce the operators

$$M_A = P + (\Lambda\Pi + K_{P,Q_A})A$$

where P is an arbitrary continuous projector on ker L and

$$M_A^* = P_A + (\Lambda\Pi + K_{P_A,Q_A})A$$

which corresponds to the particular choice $P = P_A$ in M_A .

<u>Proposition X.5</u>. *Under assumptions above, we have*

$$I - M_A = (I - K_{P,Q_A}\ A)\ (I - P - \ \Lambda\Pi A) = (I - P - \ \Lambda\Pi A)\ (I - K_{P_A,Q_A} A) \quad (X.14)$$

where $I - P - \Lambda\Pi A : X \to X$ *is a topological homeomorphism, and*

$$I - M_A^{\ast} = I - M_1 - M_2$$

where

$$I - M_1 = I - P_A - K_{P_A,Q_A} A\ , \quad M_2 = \Lambda\Pi A$$

are such that

$$\text{Im } (I - M_1) \quad \ker P_A\ , \ M_2(\text{Im } P_A) = \text{Im } P_A\ ,$$

$$M_2(\ker P_A) = (I - M_1)\ (\text{Im } P_A) = \{0\}.$$

Proof. Follows at once from the properties of P_A and Q_A and simple computations.

Let us remark here that the fact that (H_6') implies (H_6) follows easily from relation (X.14) applied to μA instead of A and the spectral properties of linear compact mappings. We shall write, for each $\mu \in R$,

$$M_A(\mu) = P + \mu(\Lambda\Pi + K_{P,Q_A})A\ ,$$

$$M_A^{\ast}(\mu) = P_A + \mu(\Lambda\Pi + K_{P_A,Q_A})A$$

It is possible to show by an example that (H_6) does not imply (H_6'). By a reasoning analogous to the one used in Proposition X.2, it is possible to prove that, if (H_6) holds, (H_6') is satisfied if and only if 0 is an isolated characteristic value for (L,A) with a multiplicity equal to dim ker L or is a regular covalue.

Proposition X.6. *If the conditions above hold, then, for each* $\mu_1,\mu_2 \in R$, *one has*

$$(I - M_A^{\ast}(\mu_1))\ (I - M_A^{\ast}(\mu_2)) = (I - M_A^{\ast}(\mu_2))\ (I - M_A^{\ast}(\mu_1)). \qquad (X.15)$$

Proof. It suffices to compute the left-hand member of (X.15) and, using properties of P_A and Q_A , to note that it is symmetric with respect to μ_1 and μ_2 .

Proposition X.7. *Under assumptions above, one has*

i) $I - M_A(\mu) = (I - P + P_A)(I - M_A^{\ast}\ (\mu))$

ii) *for each integer* $n > 0$ *and each* $\mu \neq 0$,

$$\text{Im}(I - M_A(\mu))^n = \text{Im } (I - M_A^{\ast}(\mu))^n\ .$$

Proof. Relation (i) follows at once from the fact, proved in chapter I, that

$$K_{P_A,Q_A} = (I - P_A)K_{P,Q_A}\ .$$

On the other hand, it is not difficult to check that $I - P + P_A$ is a topological homeomorphism which leaves invariant every subspace of X of the form $\text{Im } P \oplus Y$, with Y a vector subspace of $\ker P_A$. By Proposition X.5, $\operatorname{Im} P_A$ and $\ker P_A$ are invariants for $I - M_A^*(\mu)$ and, for each $\mu \neq 0$,

$$\operatorname{Im}(I - M_A(\mu)) = \operatorname{Im} P_A \oplus (I - M_A(\mu))(\ker P_A) .$$

Hence the successive iterates of $I - M_A^*(\mu)$ are of the form $\operatorname{Im} P_A \oplus Y$, with $Y \subset \ker P_A$, when $\mu \neq 0$. Lastly,

$$\begin{aligned}\operatorname{Im}(I - M_A(\mu))^n &= (I - M_A(\mu))^{n-1}(I - P + P_A)\operatorname{Im}(I - M_A^*(\mu)) \\ &= (I - M_A(\mu))^{n-1}\operatorname{Im}(I - M_A^*(\mu)) = \dots = \\ &= \operatorname{Im}(I - M_A^*(\mu))^n .\end{aligned}$$

We can now state and prove the following basic

Theorem X.4. *Under assumptions above, if* μ *is a characteristic value for* (L,A), *then*

$$\beta(\mu) = \dim \ker (I - M_A(\mu))^{n(\mu)}$$

where $n(\mu)$ *is the smallest integer such that*

$$\ker(I - M_A(\mu))^{n+1} = \ker(I - M_A(\mu))^n .$$

Proof. If $\mu = 0$, the result follows from a remark made above and the fact that $I - M_A(0) = I - P$. Let us suppose that $\mu \neq 0$ and let μ_0 be a regular covalue for (L,A). By (X.12) we have

$$I - (\mu - \mu_0)A_0 = (I - M_A^*(\mu_0))^{-1}(I - M_A^*(\mu)) .$$

Now using (X.15) we get

$$(I - M_A^*(\mu_0))^{-1}(I - M_A^*(\mu)) = (I - M_A^*(\mu))(I - M_A^*(\mu_0))^{-1} .$$

Therefore, for each positive integer n,

$$(I - (\mu - \mu_0)A_0)^n = (I - M_A^*(\mu_0))^{-n}(I - M_A^*(\mu))^n ,$$

and then the smallest integer n such that

$$\ker (I - (\mu - \mu_0)A_0)^{n+1} = \ker (I - (\mu - \mu_0)A_0)^n$$

is necessarily equal to the number $n(\mu)$ defined above, and

$$\beta(\mu) = \dim \ker (I - M_A^*(\mu))^{n(\mu)} .$$

Now, by Riesz theory, we have

$$\begin{aligned}X &= \operatorname{Im}(I - M_A^*(\mu))^{n(\mu)} \oplus \ker (I - M_A^*(\mu))^{n(\mu)} \\ &= \operatorname{Im}(I - M_A(\mu))^{n(\mu)} \oplus \ker (I - M_A(\mu))^{n(\mu)}\end{aligned}$$

which, by Proposition X.7, implies that

$$\dim \ker (I - M_A^*(\mu))^{n(\mu)} = \dim \ker (I - M_A(\mu))^{n(\mu)}$$

and achieves the proof.

Remark. It can be proved that, if $\mu \neq 0$, $\beta(\mu)$ is still equal to the (usual) multiplicity of μ as characteristic value of the linear compact mapping $K_{P,Q_A}A$, with P any projector on ker L, but that the result is false if Q_A is replaced by an arbitrary projector Q such that $\ker Q = \operatorname{Im} L$.

4. Beyong their own interest the results of the preceding section appear to be very useful in coincidence degree or index theory. In this section X and Z will again be normed vector spaces on the real field and (L,A) a pair of linear mappings verifying throughout the assumptions (H_1) and (H_5) above.

If $\mu \in R$ is not a characteristic value for (L,A), $x = 0$ is a isolated zero of $L - \mu A$ (and a isolated fixed point of $M(\mu)$) and the coincidence index

$$i\,[\,(L,\mu A),\,0] = i(\mu)$$

of L and μA at zero is well defined and equal to $i_{LS}[I - M(\mu),\,0]$ where i_{LS} means the Leray-Schauder index and where $M(\mu)$ is defined in (X.8)

Theorem X.5. *If assumptions* (H_1), (H_5), (H_6) *hold for* (L,A) *and if* μ_1, μ_2, *with* $\mu_1 < \mu_2$ *are not characteristic values for* (L,A), *then*

$$i(\mu_1) = (-1)^{\beta}\, i(\mu_2)$$

where β *is the sum of the multiplicities of the characteristic values for* (L,A) *lying in the interval* $[\mu_1,\mu_2]$.

Proof. By our assumptions, we can take for spectral operator

$$A_1 = (L - \mu_1 A)^{-1} A$$

and we deduce, from (X.12)

$$I - M(\mu_2) = [I - M(\mu_1)][I - (\mu_2 - \mu_1)A_1],$$

where $I - M(\mu_i)$ $(i = 1,2)$ are linear homeomorphisms. Using the Leray's product theorem, we obtain

$$i_{LS}[I - M(\mu_2),\,0] = i_{LS}[I - M(\mu_1),0]\; i_{LS}[I - (\mu_2 - \mu_1)A_1,\,0]$$

all the indices being well-defined because $I - (\mu_2 - \mu_1)A_1$ is also a linear homeomorphism. Since the sum of the multiplicities, in the classical sense, of the characteristic values of A_1 situated in $[0,\mu_2 - \mu_1]$ is equal, by Definition X.4, to the sum of the multiplicities of the characteristic values for (L,A) situated in $[\mu_1,\mu_2]$, the result follows immediately from the above equality, the Leray-Schauder formula for the index of linear compact mappings and the definition of the coincidence index.

An interesting special case of Theorem X.5 is the following

Corollary X.1. *If the pair* (L,A) *satisfies conditions* (H_1), (H_5), (H_6), *if* μ *is the only characteristic value for* (L,A) *situated in* $[\mu - \varepsilon, \mu + \varepsilon]$ $(\varepsilon > 0)$ *and if, for each* $x \in \ker (L - \mu A)\setminus\{0\}$, *one has*

$$Ax \notin \operatorname{Im} (L - \mu A) ,$$

then

$$i(\mu - \varepsilon) = (-1)^{\dim \ker (L - \mu A)} i(\mu + \varepsilon) .$$

Proof. The result follows immediately from Proposition X.2 and Theorem X.5, if we note that assumption (H_6) implies that

$$\ker L \cap \ker A = \{0\} ,$$

which in turn gives

$$\ker (L - \mu A) \cap \ker A = 0$$

for every $\mu \in R$, and hence the condition of Proposition X.2 can be written

$$Ax \notin \operatorname{Im} (L - \mu A)$$

for each $x \in \ker (L - \mu A)\setminus\{0\}$.

5. Let $L : \operatorname{dom} L \to Z$ satisfy (H_1), Ω be an open bounded neighbourhood of the origin in X,

$$N : R \times \overline{\Omega} \to Z, \ (\mu,x) \mapsto N(\mu,x)$$

a mapping L-compact on bounded sets of $R \times \overline{\Omega}$ and such that, for each $\mu \in R$,

$$N(\mu,0) = 0.$$

Hence, for each $\mu \in R$, $x = 0$ is a solution of the equation

$$Lx = N(\mu,x) \qquad (X.16)$$

and the following definition is classical.

Definition X.5. The point $(\mu_0,0)$ of the line $d = \{(\mu,0) \in R \times Y \mid \mu \in R\}$ will be said a *bifurcation point for the solution of* (X.16) *with respect to* d if every neighbourhood of $(\mu_0,0)$ in $R \times \overline{\Omega}$ contains at least one solution (μ,x) of (X.16) distinct of $(\mu,0)$. More briefly we shall say that μ_0 is a *bifurcation point for* (L,N).

Lemma X.5. *If* (L,N) *satisfies conditions above and if* $[\mu_1,\mu_2]$ *contains no bifurcation point for* (L,N), *then there exists* $\delta > 0$ *such that, for each* $\mu \in [\mu_1,\mu_2]$ *and each* $x \in \overline{\Omega} \cap B_\delta(0)$

$$Lx = N(\mu,x) \Rightarrow x = 0 .$$

Proof. Let us first note that it follows from the L-compactness of N that the set

$$\{(\mu,x) \mid \mu \in [\mu_1, \mu_2], x \in \overline{\Omega}, Lx = N(\mu,x)\}$$

is compact in R x X. Suppose now that Lemma X.5 is false. Then, for each $n \in N^*$, there exists $\tilde{\mu}_n \in [\mu_1,\mu_2]$ and $x_n \in \overline{\Omega} \cap B_{\frac{1}{n}}(0)$ such that

$$Lx_n = N(\tilde{\mu}_n,x_n) \text{ and } x_n \neq 0 . \tag{X.17}$$

Taking, if necessary, a subsequence, we can suppose that $(\tilde{\mu}_n,x_n)$ converges to (μ_0,x_0), and, necessarily, we will have $\mu_0 \in [\mu_1,\mu_2]$ and $x_0 = 0$. But then, by (X.17) and the above definition, μ_0 is a bifurcation point for (L,N), a contradiction.

Theorem X.6. *If* (L,N) *satisfies conditions above, if* $\mu_1,\mu_2 \in R$ $(\mu_1 < \mu_2)$ *are such that*

$$i(\mu_j) = i [(L,N(.,\mu_j)), 0] , j = 1,2,$$

are defined and if

$$i(\mu_1) \neq i(\mu_2) ,$$

then there exists $\mu_0 \in [\mu_1,\mu_2]$ *such that* μ_0 *is a bifurcation point for* (L,N).

Proof. Because of $i(\mu_1)$ and $i(\mu_2)$ are defined, there exists $\delta_0 > 0$ such that $x = 0$ is the unique solution of equations

$$Lx = N(\mu_j,x) , j = 1,2,$$

contained in $B_\delta(0)$. Suppose now that $[\mu_1, \mu_2]$ contains no bifurcation point for (L,N) and let $\delta_1 > 0$ be the number given by Lemma X.5. Then, for each $\delta \in]0, \min(\delta_0,\delta_1)]$, each $\lambda \in [0,1]$ and each $x \in B_\delta(0)$,

$$Lx = N(\lambda\mu_2 + (1-\lambda)\mu_1,x) \Rightarrow x = 0 .$$

Therefore, by the invariance of coincidence degree with respect to L-compact homotopies,

$$i(\mu_1) = d [(L,N(\mu_1,.),B(0,\delta)] = d [(L,N(\mu_2,.),B(0,\delta)] = i(\mu_2),$$

a contradiction.

Theorem X.6. is very general but difficult to apply because of the necessity of estimating $i(\mu_1)$ and $i(\mu_2)$. To obtain more explicit criteria, we shall suppose that

$$N(\mu,x) = \mu A + R(\mu,x)$$

where $A : X \to Z$ is linear and L-compact, and $R : R \times \overline{\Omega} \to Z$ is L-compact on bounded subsets of R x and such that

$$R(\mu,x) = o_L(\|x\|)$$

uniformly in μ on compact intervals. A necessary condition for the existence of a bifurcation point is given by the following

Theorem X.7. *If* (L,N) *satisfies assumptions above and if* μ_o *is a bifurcation point for* (L,N) *then* μ_o *is a characteristic value for* (L,A).

Proof. Suppose that μ_o is not a characteristic value for (L,A). Then,

$$L_o = L - \mu_o A$$

has an inverse L_o^{-1} and it follows from Lemma X.1 that the mappings

$$(\mu,x) \mapsto (\mu-\mu_o)L_o^{-1}A$$

$$(\mu,x) \mapsto L_o^{-1} R(\mu,x)$$

are compact on bounded sets of $R \times \overline{\Omega}$ and that

$$L_o^{-1}R(\mu,x) = o(\|x\|)$$

uniformly in μ on compact intervals. Now, for $x \in (\mathrm{dom}\, L \cap \overline{\Omega}) \setminus \{0\}$,

$$\|L_o^{-1}(L\,x - N(\mu,x))\| = \|L_o^{-1}(L_o x - (\mu-\mu_o)Ax-R(\mu,x)\|$$

$$\geq \|x\| - |\mu-\mu_o|\,\|L_o^{-1}\, Ax\| - \|L_o^{-1}R(\mu,x)\|$$

$$\geq \|x\|[\,1 - |\mu-\mu_o|\,\|L_o^{-1}A\| - \|x\|^{-1}\|L_o^{-1}R(\mu,x)\|] \geq \|x\|/3$$

if first μ is in $[\mu_o-\delta_1, \mu_o+\delta_1]$ with $\delta_1 > 0$ such that $\delta_1\|L_o^{-1}.A\| \leq 1/3$ and then $x \in B_{\rho_1}(0)$ with $\rho_1 > 0$ such that $\|x\|^{-1}\|L_o^{-1}R(\mu,x)\| \leq 1/3$ for each $(\mu,x) \in [\mu_o-\delta_1, \mu_o+\delta_1] \times B_{\rho_1}(0)$. Therefore, μ_o cannot be a bifurcation point for (L,N), a contradiction.

We give now a sufficient condition for the existence of a bifurcation point.

Theorem X.8. *If* (L,N) *satisfies the assumptions above, if condition* (H_6) *holds for* (L,A) *and if* μ_o *is a characteristic value for* (L,A) *of odd multiplicity, then* μ_o *is a bifurcation point for* (L,N).

Proof. First, μ_o is isolated and hence, for sufficiently small $\varepsilon > 0$, it follows from Theorem X.5 and Theorem X.2 that

$$i(\mu_o - \varepsilon) = i\,[\,(L,\ (\mu_o - \varepsilon)A),0] = (-1)^{\beta_o}\, i[\,(L,\ (\mu_o + \varepsilon)A),0] = (-1)^{\beta_o}\, i(\mu_o + \varepsilon)$$

where β_o is the multiplicity of μ_o. The result follows then from Theorem X.6 and the oddness of β_o.

A result much more easy to apply is given by the following

Corollary X.2. *Under conditions above for* (L,N) *if condition* (H_6) *holds for* (L,A) *and if* μ_o *is a characteristic value of* (L,A) *such that* :

(i) $\forall\ x \in \ker (L - \mu_o A)\setminus\{0\}$, $Ax \in \mathrm{Im}(L - \mu_o A)$;

(ii) $\dim \ker (L - \mu_o A)$ *is odd,*

then μ_o *is a bifurcation point for* (L,N).

Proof. The result follows easily from Theorem X.8, Proposition X.2 and the remark following this Proposition.

6. As an application of the bifurcation theory given in section we shall consider the nonlinear Steklov problem for a elliptic equation.

Let $D \subset R^n$ be a bounded open set with boundary Γ Hölder continuous of class $C^{1+\alpha}$ for some $\alpha > 0$. If $C(\overline{D})$(resp. $C(\Gamma)$) denotes the Banach space of real continuous functions on $\overline{D}$ (resp. Γ) with the uniform norm, the restriction $\tilde{x}$ to Γ of any element x of $C(\overline{D})$ is an element of $C(\Gamma)$. If $C^j(D)$ is the set of real functions of class C^j in D, let us consider the mapping

$$L \ : C(D) \cap C^2(D) \to C(D)\ ,\ x \mapsto \sum_{i,j=1}^{n} \frac{\partial}{\partial w_i}\left(a_{ij}(w)\frac{\partial x}{\partial w_j}\right)$$

where $w = (w_1,\dots,w_n) \in \overline{D}$, the real functions a_{ij} are continuous on $\overline{D}$, $a_{ij} = a_{ji}$ $(i,j = 1,2,\dots,n)$, the first partial derivatives of the a_{ij} are uniformly Hölder continuous on D and there exists $\gamma > 0$ such that, for each $w \in D$ and each $y \in R^n$,

$$\sum_{i,j=1}^{n} a_{ij}(w)y_i y_j \geq \gamma \left(\sum_{i=1}^{n} y_i^2\right).$$

Thus L is a formally self-adjoint elliptic differential operator. If

$$h : \Gamma \times R \times R \to R\ ,\ (w,\xi,\mu) \to h(w,\xi,\mu)$$

is continuous and such that

$$|\xi|^{-1}|h(w,\xi,\mu)| \to 0 \text{ if } |\xi| \to 0$$

uniformly in $w \in \overline{D}$ and μ on compact intervals, the nonlinear Steklov problem consists in determining the solutions $(\mu,x) \in R \times (C^2(D) \cap C(\overline{D}))$ of the equations

$$\begin{aligned} &L\,x(w) = 0\ ,\ w \in D \\ &\frac{\partial x}{\partial \nu}(w) = \mu u(w) + h(w,u(w),\mu)\ ,\ w \in \Gamma, \end{aligned} \tag{X.18}$$

where

$$\frac{\partial x}{\partial \nu}(w) = \sum_{i,j=1}^{n} a_{ij}(w) n_j(w) \frac{\partial u}{\partial w_i}$$

with $n(w) = (n_1(w),\ldots,n_n(w))$ is the unit exterior normal to Γ at w.

If we take

$$X = C(\overline{D}),\ Z = C(\Gamma),\ \text{dom}\ L = \{x \in C^1(\overline{D}) \cap C^2(D) : \mathcal{L}\,x = 0\}$$

$$L : \text{dom}\ L \to Z,\ x \to \frac{\partial x}{\partial \nu},\ A : X \to Z,\ x \to \tilde{x}, \tag{X.19}$$

$$R : R \times X \to Z,\ (\mu,x) \to h(.,\tilde{x}(.),\mu),$$

then the nonlinear Steklov problem is clearly equivalent to the operator equation in dom L

$$Lx = \mu Ax + R(\mu,x) \tag{X.20}$$

which has, for each $\mu \in R$, the trivial solution $(\mu,0)$.

Theorem X.9. *Under assumptions listed above, each characteristic value* μ *for* (L,A) *with* L *and* A *defined in* (X.19), *such that* dim ker $(L - \mu A)$ *is an odd number is a bifurcation point for the nonlinear Steklov problem* (X.18)

Proof. It follows from the study of the linear Neumann problem that

(i) $\ker L = \{x \in \text{dom}\ L : x(w) \text{ is constant},\ w \in \overline{D}\}$

(ii) $\text{Im}\ L = \{y \in Z : \int_\Gamma y(w)dS_w = 0\}$,

where dS_w is the measure element on Γ. Hence L is a Fredholm mapping of index zero and if we define $P : X \to X$ by

$$(Px)(w) = (\text{meas}\ \Gamma)^{-1} \int_\Gamma x(z)dS_z,\quad w \in \overline{D},$$

then P is a continuous projector such that Im P = ker L and its restriction Q to Z is a projector such that ker Q = Im L. It follows also from the study of the linea Neumann problem that

$$(L_P^{-1}y)(w) = \int_\Gamma N(w,z)y(z)dS_z,\quad w \in \overline{D},$$

where the Neumann kernel $N(w,z)$ has regularity properties making L_P^{-1} a compact mapping. This easily implies that A and R are L-compact on bounded sets, and hence all the basic assumptions are satisfied for the pair (L,A+R). Lastly, the assumptic imply that the linear problem

$$Lx - \mu Ax = y$$

has a solution if and only if

$$\int_{\Gamma} y(w)x(w)dS_w = 0 . \qquad (X.21)$$

Hence $x(w) = 0$ for each $w \in \Gamma$ and, because of $x \in \operatorname{dom} L$, the maximum principle implies that $x(w) = 0$, $w \in \overline{D}$, a contradiction. Because of $\mu = 0$ is a characteristic value, assumption (H_6'), and hence assumption (H_6) hold. Theorem X.9 follows then directly from Corollary X.2.

7. Bibliographical notes about Chapter X

The concept of coincidence index (Definition X.1) is introduced in Laloux and Mawhin(*Trans. Amer. Math. Soc.*, to appear) as well as the content of section 1 and Theorems X.1 and X.2 of section 2. Lemma X.1 is proved in more generality in Laloux ("Equ. differ. et fonctionnelles non linéaires", Hermann, 1973, 110-121). Theorem X.3 is due to Laloux (*Ann. Soc. Sci. Bruxelles* 88 (1974) 176-182) and generalizes earlier results of Melamed (*Dokl. Ak. Nauk SSSR* 126 (1959) 501-504; *Sibirsky Math. Zt.* 2(1961) 413-427). The special case where k=0 is given in Laloux and Mawhin (*Trans. Amer. Math. Soc.*, to appear) where it is shown how this results generalizes a theorem of Krasnosel'skii ("Topological methods in the theory of nonlinear integral equations", Pergamon, 1963, 216-223). For applications of Theorem X.3 to the existence of periodic solutions of ordinary differential equations with a small parameter, extending a result of Halanay (*Atti. Accad. Naz. Lincei, Rend. Cl. Sci. Fis., Mat., Natur.* (8) 22 (1957) 30-32) see Mawhin ("Intern. Conf. Diff. Equ.", Acad. Press, 1975, 537-556 and *Bol. Un. Mat. Ital.*, to appear). The concepts of characteristic values and regular covalues for (L,A) are given in various degrees of generality in Laloux ("Equ. differ. et fonctionnelles non linéaires", Hermann, 1973, 110-121) and Laloux and Mawhin (*Trans. Amer. Math. Soc.*, to appear), the treatment of section 3 being the one in Laloux and Mawhin(Multiplicity, Leray-Schauder formula and bifurcation, *to appear*) where Definition X.4 of multiplicity is given. Under the assumption (H_6') the formula given in Theorem X.4 had precedingly adapted to define the multiplicity in Laloux and Mawhin (*Trans. Amer. Math. Soc.*, to appear). Theorem X.5 is given in Laloux and Mawhin (Multiplicity, Leray-Schauder formula and bifurcations, *to appear*) and generalizes a well known formula of Leray and Schauder (*Ann. Ecole Norm. Sup.* 51(1934) 45-78). Other forms of Theorem X.5 are given in Laloux and Mawhin (*Trans. Amer. Math. Soc.*, to appear). The concept of bifurcation point can be found for example in Krasnosel'skii ("Topological method in the theory of nonlinear integral equations", Pergamon, 1963, ch. IV). Lemma X.5 and Theorem X.6 and X.7 are due to Laloux ("Equ. differ. et fonctionnelles non linéaires", Hermann, 1973, 110-121) and generalize results of Krasnosel'skii (op. cit.). Theorem X.3 and Corollary X.2 are due to Laloux and

Mawhin (Multiplicity, Leray-Schauder formula and bifurcations, to appear) and also generalize Krasnosel'skii results. All those results are local in nature and extensions to the frame of coincidence degree of the global bifurcation theorem of Rabinowitz (*J. Functional Anal.* 7 (1971) 487-513) can be found in Laloux ("Indice de coïncidence et bifurcations", Thèse de Doctorat, Louvain, 1974) where further results and applications in the line of this Chapter can be found. Theorem X.9 was first proved by Stuart and Toland (*J. Differential Equations* 15 (1974) 247-268) using another and lenghtier argument. The proof given here can be found in Laloux and Mawhin ("Multiplicity, Leray-Schauder formula and bifurcations", to appear). For results about the linear Neumann problem which are used in section 6, see Cushing (*Arch. Rat. Mech. Anal.* 42 (1971) 63-76) and Miranda ("Partial Differential Equations of elliptic type", Springer 1970). For the whole chapter see also the survey paper by Mawhin (*Berichte Geselsch. Mathem. Datenverarbeitung Bonn* 103 (1975) 7-22).

XI. COINCIDENCE DEGREE FOR k-SET CONTRACTIVE PERTURBATIONS OF LINEAR FREDHOLM MAPPINGS

1. If Y is a metric space and B a subset of Y, the *(Kuratovski) measure of noncompactness* $\alpha(B)$ *of* B is defined by

$$\alpha(B) = \inf \{d > 0 : B \text{ has a finite cover by sets of diameter smaller than } d\}.$$

If Y_1 and Y_2 are metric spaces, a continuous mapping $f : Y_1 \to Y_2$ will said to be a *k-set contraction* if there exist a nonnegative real k such that, for each bounded $B \subset Y_1$, on has

$$\alpha(f(B)) \leqslant k\, \alpha(B) .$$

We list the main properties of k-set contractions :

a. If Y_3 is a metric space and $f : Y_1 \to Y_2$, $g : Y_2 \to Y_3$ are respectively k_1- and k_2- set contractions, then $gf : Y_1 \to Y_3$ is a k_1k_2-set contraction

b. If E is a normed space and $f : Y_1 \to E$, $g : Y_1 \to E$ are respectively k_1- and k_2-set contractions, then $f + g : Y_1 \to E$ is a $(k_1 + k_2)$-set contraction.

c. If $f : Y_1 \to E$ is a k-set contraction and $\lambda \in R$, then λf is a $|\lambda|$k-set contraction.

d. Let $V : Y_1 \times Y_1 \to Y_2$ continuous, such that for each $y \in Y_1$, $V(.,y) : Y_1 \to Y_2$ is Lipschitzian with a constant k independent of y and such that for each bounded $A \subset Y_1$, the mapping $y \to V(.,y)$ is a compact mapping between A and the metric space $C(A,Y_2)$ of continuous, bounded mappings from A into Y_2 with the uniform topology. Then if we define $f : Y_1 \to Y_2$ by

$$f(x) = V(x,x) ,$$

f is a k-set contraction.

As special cases we get

e. If $f : Y_1 \to Y_2$ is Lipschitzian of constant k (resp. compact on bounded sets of Y_1) then f is a k-set contraction (resp. a 0-set contraction)

2. Let now $\Omega \subset X$, X a Banach space, be open and bounded and $f : \overline{\Omega} \to X$ a k-set contraction with $k < 1$ such that

$$0 \notin (I - f)(\partial\Omega) .$$

Then one can define an integer, the degree of $I - f$ with respect to Ω and 0,

$$d(I - f, \Omega, 0)$$

which has the following basic properties of Leray-Schauder degree to which it reduces when f is compact.

(a) If $d(I - f, \Omega, 0) \neq 0$, then $0 \in (I - f)(\Omega)$.

(b) If Ω_1 and Ω_2 are disjoint open subsets of Ω such that $0 \notin (I - f)[\overline{\Omega} \setminus (\Omega_1 \cup \Omega_2)]$, then

$$d(I - f, \Omega, 0) = d(I - f, \Omega_1, 0) + d(I - f, \Omega_2, 0) .$$

(c) If $\Omega \ni 0$ and is symmetric with respect to 0 and if $f(-x) = - f(x)$ on $\partial\Omega$, then

$$d(I - f, \Omega, 0) = 1 \pmod 2$$

(d) If $F : \overline{\Sigma} \to X$ with $\Sigma \subset X \times [0,1]$ open bounded, is continuous, such that

$$x - F(x,\lambda) \neq 0$$

for each $x \in \partial\Sigma$ and if, for each bounded $B \subset X$, and some $k < 1$,

$$\alpha [F(\Sigma \cap (B\times[0,1]))] \leqslant k\, \alpha(B) ,$$

then

$$d [I - F(.,\lambda), \Sigma_\lambda, 0]$$

is independent of λ in $[0,1]$ with $\Sigma_\lambda = \{x \in X : (x,\lambda) \in \Sigma\}$.

3. Let now X and Z be real Banach spaces, $L : \text{dom } L \subset X \to Z$ be a linear Fredholm mapping of index zero, for which notations of chapter 1 will be conserved and $N : \overline{\Omega} \to Z$ be a mapping.

Definition XI.1. N will be said a *L-k-set contraction* if :

a. $\Pi N : \overline{\Omega} \to \text{coker } L$ is continuous and $\Pi N(\overline{\Omega})$ bounded.

b. $K_{P,Q} N : \overline{\Omega} \to X$ is a k-set contraction.

This definition is justified by the following

emma XI.1. *Definition* XI.1 *is independent upon the choice of projectors* P *and*
Q

roof. Let $\tilde{P}$, $\tilde{Q}$ other projectors such that $\operatorname{Im}\tilde{P} = \ker L$, $\operatorname{Im} L = \ker \tilde{Q}$.
hen, using (I.7),

$$K_{\tilde{P},\tilde{Q}} N = (I - \tilde{P})K_{P,Q} N + (I - \tilde{P}) K_P(Q - \tilde{Q})N .$$

y properties (a) to (e) of section 1 and compactness of $\tilde{P}$, $(I - \tilde{P})K_{P,Q} N$ is a
-set contraction. Now $(I - \tilde{P})K_P(Q - \tilde{Q})N$ is continuous, bounded and finite-
imensional, thus compact, and therefore is a 0-set contraction. This implies
hat $K_{\tilde{P},\tilde{Q}} N$ is a k-set contraction.

. Let now L and N be like above and assume that $k < 1$ and that

$$0 \notin (L - N)(\operatorname{dom} L \cap \partial\Omega).$$

he notations being those of Chapter III,

$$M_\Lambda = P + (\Lambda\Pi + K_{P,Q})N$$

s a k-set contraction and hence

$$d\,[I - M_\Lambda, \Omega, 0]$$

s defined. We have the following

emma XI.2. $d\,[I - M_\Lambda, \Omega, 0]$ *depends only upon* L, N, Ω *and the homotopy class*
f Λ *in* L_L.

roof. Using notations of Proposition III.6, let us consider the mapping M defined
n $\bar{\Omega} \times [0,1]$ by

$$M_\Lambda(x,\lambda) = P(\lambda)x + \Lambda\,[\Pi x,\lambda] + K_{P(\lambda),Q(\lambda)} Nx .$$

s in Proposition III.6,

$$x \neq M_\Lambda(x,\lambda)$$

or all $x \in \partial\Omega$ and $\lambda \in [0,1]$. We have only to show that, for each bounded
$\subset \bar{\Omega}$,

$$\alpha\,[M_\Lambda(B\times[0,1])] \leq k\,\alpha(B) .$$

Now,

$$\alpha\,[M_\Lambda(B\times[0,1])] \leq \alpha(P\,[B\times[0,1]]) + \alpha\,\{\Lambda\,[\Pi N(B)\times[0,1]]\} + S(B\times[0,1]) \qquad (XI.1)$$

if

$$S(x,\lambda) = K_{P(\lambda),Q(\lambda)}\,Nx$$

$$= ((1-\lambda)K_P + \lambda K_{P'})(I-(1-\lambda)Q - \lambda\,Q')Nx$$

As $P\,[B\times[0,1]]$ and $\Lambda\,[\Pi N(B)\times[0,1]]$ are bounded and finite-dimensional, they are relatively compact and hence the two first terms in (XI.1) are zero.

Now,

$$\alpha\,[\bigcup_{\lambda\in[0,1]} K_{P(\lambda),Q(\lambda)}\,N(B)\,]$$

$$\leq \alpha\,\{\bigcup_{\lambda\in[0,1]}[(1-\lambda)K_P + \lambda K_{P'}]\bigcup_{\lambda'\in[0,1]}(I-Q(\lambda'))\,N(B)\}$$

$$\leq \alpha\,\{co\,[K_P(\bigcup_{\lambda'\in[0,1]}(I-Q(\lambda'))N(B)) \cup K_{P'}(\bigcup_{\lambda'\in[0,1]}(I-Q(\lambda'))N(B))]\}$$

$$\leq \max\,\{\alpha(K_P\bigcup_{\lambda'\in[0,1]}(I-Q(\lambda'))N(B)),\ \alpha(K_{P'}(\bigcup_{\lambda'\in[0,1]}(I-Q(\lambda'))N(B))\}$$

$$\leq \max\,\{\alpha(co(K_{P,Q}\,N(B)\cup K_{P,Q'}\,N(B))),\ \alpha(co(K_{P',Q}\,N(B)\cup K_{P',Q'}\,N(B)))\}$$

$$\leq k\,\alpha(B)\ .$$

The proof nows ends like in Proposition III.6 using property (d) of the degree.

Proceeding now like in Chapter III we can introduce and justify the following

Definition XI.2. If conditions listed in section 3 and 4 are satisfied for L,N and Ω , the *coincidence degree* $d\,[(L,N),\,\Omega\,]$ *of* L *and* N *in* Ω is the integer

$$d\,[\,(L,N),\,\Omega\,] = d\,[\,I - M_\Lambda\,,\,\Omega,\,0\,]$$

where the right-hand member is the degree for k-set contractive perturbation of identity and Λ is any orientations preserving isomorphism from coker L into ker L.

Now using properties of $d\,[I - M_\Lambda,\,\Omega,\,0]$ listed above, we immediately get the properties of $d[(L,N),\Omega]$ listed in Theorem III.1 for our new setting and Theorem III.2 if we replace the assumption of L-compactness of $\tilde{N}$ by $\Pi\tilde{N}$ continuous and bounded and

$$\alpha(K_{P,Q}\tilde{N}(B\times[0,1])) \leq k\,\alpha(B)$$

for some $k < 1$ and each bounded $B \subset \Omega$, an assumption which still does not depend upon the choice of P,Q.

Also, the generalized continuation theorem IV.1 is easily formulated and similarly proved in the context of L-k-set contractions.

Other measures of noncompactness have been defined and are useful and in particular the *(Hausdorff) - ball measure of noncompactness* $X_M(A)$ of the bounded set A in the metric space M which is defined by

$X_M(A) = \inf\{\varepsilon > 0$: A can be covered by a finite number of balls of radius ε and center in $M\}$.

Then if Y_1 and Y_2 are metric spaces and $f : Y_1 \to Y_2$ is continuous and such that there exists a real $k \geq 0$ such that for each bounded set $B \subset Y_1$, one has

$$X_{Y_2}(f(B)) \leq k\, X_{Y_1}(B) ,$$

f will be called a k-*ball contraction*. As a degree theory exists, with the properties (a) - (b) - (c) - (d) of section 2 for k-ball contractive perturbations of identity in a Banach space when $k < 1$, a coincidence degree theory can be built similarly for couples (L,N) with L like in section 3 and N a L-k-ball contraction.

6. It is interesting for application to look for sufficient conditions upon L and N which insure that N is L-k-set contractive.

Let us recall that a $\emptyset_+$*-operator* $L : \operatorname{dom} L \subset X \to Z$ is a closed linear operator such that $\dim \ker L < \infty$ and Im L is closed.

Proposition XI.1. L *is a* $\emptyset_+$*-operator if and only if there exists* $r > 0$ *such that, for each bounded* $B \subset \operatorname{dom} L$,

$$\alpha(L(B)) \geq r\,\alpha(B) . \qquad \text{(XI.2)}$$

Proof. The assertion is trivial if dom L or Z is finite-dimensional so we can assume that both are infinite dimensional. Let us first prove the necessity. Let P be a projector onto ker L. Then

$$X = \ker L \oplus \ker P$$

and let $L_P = L|\ker P \cap \mathrm{dom}\, L$. Then L_P is one-to-one, closed and $\mathrm{Im}\, L_P = \mathrm{Im}\, L$ is closed which implies by the closed graph theorem that $K_P = L_P^{-1} : \mathrm{Im}\, L \to X$ is continuous. Therefore we can find $r > 0$ such that, for each $x, y \in \mathrm{dom}\, L \cap \ker P$,

$$|L_P x - L_P y| \geq r|x - y| . \tag{XI.3}$$

Now if $\tilde{B} \subset \ker P \cap \mathrm{dom}\, L$ is bounded and if $\{D_1, \ldots, D_n\}$ is a covering of $L(B)$ with $\mathrm{diam}\, D_i \leq d$ for $1 \leq i \leq n$, it follows from (XI.3) that

$$r\, \mathrm{diam}(L_P^{-1}(D_i)) \leq d\ ,\ 1 \leq i \leq n$$

and therefore,

$$r\, \alpha(\tilde{B}) \leq \alpha(L_P(\tilde{B}))$$

for each bounded $\tilde{B} \subset \ker P \cap \mathrm{dom}\, L$. Now, if B is bounded in $\mathrm{dom}\, L$,

$$\begin{aligned} r\, \alpha(B) &\leq r\, \alpha\, [(I - P)(B) + P(B)] \\ &\leq r\, \alpha((I - P)(B)) \leq \alpha(L_P(I - P)(B)) \\ &\leq \alpha(L(B)) + \alpha(LP(B)) = \alpha(L(B)). \end{aligned}$$

Let us now prove the sufficiency. If B_1 denotes the unit ball in X and $B = B_1 \cap \ker L$, one has

$$r\, \alpha(B) \leq \alpha(\{0\}) = 0$$

and hence B is compact which implies by Riesz theorem that $\dim \ker L < \infty$. Hence we still have $L_P : \mathrm{dom}\, L \cap \ker P \to \mathrm{Im}\, L$ is bijective and, according to the theory of Fredholm mappings, $\mathrm{Im}\, L$ will be closed if $L_P^{-1} : \mathrm{Im}\, L \to \mathrm{dom}\, L \cap \ker P$ is continuous. If not, there exists a sequence $\{x_n\}$ in $\ker P$ such that $|x_n| = 1$ and $L_P x_n \to 0$. Therefore

$$\alpha\, (\{L_P x_n : n \in N\}) = 0 \tag{XI.4}$$

and using (XI.2),

$$\alpha\, (\{x_n : n \in N\}) = 0$$

This implies that $\{x_n : n \in N\}$ is relatively compact and hence there exists a subsequence $\{x_{n'}\}$ such that $x_n \to x \in \ker P$. Clearly $|x| = 1$ and, L_P being closed, $L_P x = Lx = 0$, i.e. $x \in \ker L$, a contradiction. We are then led to the

<u>Definition XI.3</u>. If $L : \mathrm{dom}\, L \subset X \to Z$ is a ϕ_+- operator , then
$l(L) = \sup \{r \in R_+ : \text{for each bounded } B \subset \mathrm{dom}\, L,\ r\, \alpha(B) \leq \alpha(L(B))\}$.

The set used in the definition of $l(L)$ being closed, $l(L)$ is a maximum.

Proposition XI.2. *Let* $L : \operatorname{dom} L \subset X \to Z$ *be a closed Fredholm mapping of index zero and* $N : \bar{\Omega} \subset X \to Z$ *be a* k*-set contraction with*

$$0 \leq k' < l(L).$$

Then N *is a* L-k*-set contraction with constant* $k = \frac{k'}{l(L)} < 1$.

Proof. The assumption upon ΠN is clearly satisfied. Let $B \subset \bar{\Omega}$ be a bounded set. Then

$$\alpha(K_{P,Q} N(B)) = \alpha(L_P^{-1} (I - Q)N(B))$$

$$\leq [l(L)]^{-1} \alpha[(I - Q)N(B)] \leq [l(L)]^{-1} k'\alpha(B).$$

Now Proposition XI.1 can be used to obtain an useful result about linear perturbations of ϕ_+-operators that we shall give without proof.

Proposition XI.3. *Let* X,Z *be Banach spaces,* $L : \operatorname{dom} L \subset X \to Z$ *be a* ϕ_+*-operator and* $T : X \to Z$ *a linear* k*-set contraction with* $k \in [0, l(L)[$. *Then* $L + T$ *is a* ϕ_+*-operator and*

$$\operatorname{Ind} L = \operatorname{Ind}(L + T).$$

It is easily checked that Proposition XI.1 also holds for the ball-measure of noncompactness χ, that a number $\tilde{l}(L)$ can be associated to any ϕ_+-operator by Definition XI.3 with α replaced by χ and that Propositions XI.2 and XI.3 hold in this context.

6. We shall apply the above theory to the proof of a Fredholm alternative for some nonlinear mappings.

Definition XI.4. If X,Z are Banach spaces and $T : X \to Z$ is continuous, we shall call T *asymtotically linear* if there exists a continuous linear mapping $B : X \to Z$ such that

$$\lim_{|x| \to \infty} \frac{|Tx - Bx|}{|x|} = 0 .$$

It is seen at once that if such a linear B exists, it is unique and it is called the *asymptotic derivative* of B. Of course any asymptotically linear mapping is quasibounded. Moreover one can prove the following.

Proposition XI.4. *If* $T : X \to Z$ *is asymptotically linear and is a* k-set (ball)-*contraction, then its asymptotic derivative is a* k*-set (ball) - contraction.*

Let us prove now the following basic

<u>Lemma XI.3</u>. *Let* $L : \mathrm{dom}\, L \subset X \to Z$ *be a closed Fredholm mapping of index zero,* $T : X \to Z$ *a k-set contraction asymptotically linear with* $k \in [0, l(L)[$ *and* B *be the asymptotic derivative of* T. *Suppose that*

$$\mathrm{Im}\,(T - B) \subset \mathrm{Im}\,(L - B).$$

Then $0 \in \mathrm{Im}\,(L - T)$.

<u>Proof</u>. By Proposition XI.4, B is a k-set contraction and hence using Proposition XI.3, $L - B$ is a Fredholm operator with $\mathrm{Ind}(L - B) = \mathrm{Ind}\, L = 0$. Let therefore $S : \ker\,(L - B) \to V$ be an isomorphism with V a vector subspace such that $Z = \mathrm{Im}\,(L - B) \oplus V$. Then if $P : X \to X$ is a continuous projector onto $\ker(L - B)$, $C = SP$ will have finite range and hence is compact, which implies that $L - B - C$ is a Fredholm operator of index zero. By definition of C,

$$Lx - Bx - Cx = 0\ ,\ x \in \mathrm{dom}\, L$$

is equivalent to

$$Lx - Bx = 0,\ Cx = 0\ ,\ x \in \mathrm{dom}\, L\ ,$$

i.e. to

$$Lx - Bx = 0,\ x \in \ker P \cap \mathrm{dom}\, L$$

which implies

$$x = 0$$

by the definition of P. Thus $L - B - C$ is one-to-one and therefore is onto. By the closed graph theorem, $(L - B - C)^{-1}$ is continuous and hence there exists $m > 0$ such that, for all $x \in \mathrm{dom}\, L$,

$$|Lx - Bx - Cx| \geq m\,|x|\,.$$

By definition of B there exist $\rho > 0$ such that for all x with $|x| \geq \rho$,

$$|Tx - Bx| \leq \frac{m}{2}|x|\,.$$

Now if $\Omega = B(0,\rho)$ and $N : \overline{\Omega} \times [0,1] \to Z$ is defined by

$$N(x,\lambda) = \lambda T + (1 - \lambda)Bx + Cx$$

it is easy to check that N is a k-set contraction and if $x \in \partial\Omega$ and $\lambda \in [0,1]$,

$$|Lx - N(x,\lambda)| \geq |Lx - Bx - Cx| - |\lambda|\,|Tx - Bx|$$

$$\geq \frac{m\rho}{2}\,.$$

Hence

$$d[(L,B+C),\Omega] = d[(L,T+C),\Omega]$$

and the left-hand side is an odd number by the generalized Borsuk theorem. Thus

$$d[(L,T+C),\Omega] \neq 0$$

and there exists $x \in \operatorname{dom} L$ such that

$$Lx = Tx + Cx\ .$$

i.e.

$$Lx - Bx + Bx - Tx = Cx\ .$$

As $\operatorname{Im}(T-B) \subset \operatorname{Im}(L-B)$, both members belong to supplementary subspaces and are thus both equal to zero, which achieves the proof.

Theorem XI.1. *Under the assumptions of Lemma* XI.3,

$$\operatorname{Im}(L-T) = \operatorname{Im}(L-B).$$

Proof. If $y \in \operatorname{Im}(L-T)$, then $y = Lx - Tx = Lx - Bx + Bx - Tx$ for some $x \in \operatorname{dom} L$ and hence $y \in \operatorname{Im}(L-B)$. Now if $y \in \operatorname{Im}(L-B)$, let $\tilde{T}x = Tx + y$ for $x \in X$. Then T is a k-set contraction and

$$\tilde{T}x - Bx = Tx - Bx + y \in \operatorname{Im}(L-B)$$

and hence, by Lemma XI.3, $0 \in \operatorname{Im}(L-\tilde{T})$ i.e. $y \in \operatorname{Im}(L-T)$.

Let now $X,\tilde{X},Z,\tilde{Z}$ be Banach spaces and $<\,,\,>_1$, $<\,,\,>_2$ continuous bilinear forms on $X \times \tilde{X}$, $Z \times \tilde{Z}$. $(X,\tilde{X},<\,,\,>_1)$ or $(Z,\tilde{Z},<\,,\,>_2)$ is called a *dual system.* We shall use the following result about linear Fredholm operators.

Lemma XI.4. *Let* $(X,\tilde{X},<,>_1)$ *and* $(Z,\tilde{Z},<,>_2)$ *be dual systems*, $L : \operatorname{dom} L \subset X \to Z$, $L^t : \operatorname{dom} L^t \subset \tilde{Z} \to \tilde{X}$ *linear Fredholm mappings of index zero* with $\operatorname{dom} L = X$ *such that*

$$< Lx,y >_2 = < x,L^t y >_1$$

for all $x \in \operatorname{dom} L$, $y \in \operatorname{dom} L^t$. *Then*

(a) $\dim\ker L = \dim\ker L^t$

(b) $\operatorname{Im} L = (\ker L^t)^{\perp}$

(c) $\operatorname{Im} L^t = (\ker L)^{\perp}$.

We deduce from Theorem XI.1 and Lemma XI.4 the following

Theorem XI.2. *Let* $X,\tilde{X},Z,\tilde{Z}$, L,L^t *be like in Lemma* XI.4, $k_1 \in [0,l(L)[$, $k_2 \in [0,l(L^t)[$, $T : X \to Z$, $\tilde{T} : \tilde{Z} \to \tilde{X}$ *asymptotically linear with respective asymptotic derivatives* B,B^t *and respectively* k_1 *and* k_2*-set contractions. Suppose moreover that*

$$< Bx,y >_2 = < x,B^t y >$$

for all $x \in X$ *and* $y \in \tilde{Z}$. *Then either* $L - T$ *and* $L^t - \tilde{T}$ *are onto or* $\ker (L - B) \neq \{0\}$. *In this last case :*

(1) If $\mathrm{Im}\,(T - B) \subset (\ker (L^t - B^t))^{\perp}$ then

$$\mathrm{Im}\,(L - T) = (\ker (L^t - B^t))^{\perp} .$$

(2) If $\mathrm{Im}\,(\tilde{T} - B^t) \subset (\ker (L - B))^{\perp}$ then

$$\mathrm{Im}\,(L^t - \tilde{T}) = (\ker (L - B))^{\perp} .$$

Proof. If $\ker (L - B) = \{0\}$, then $L - B$ is onto and then the same is true for $L - T$ by Theorem XI.1. By Lemma XI.4 applied to $L - B$, $\mathrm{Im}\,(L^t - B^t) = (\ker (L - B))^{\perp} = \tilde{X}$ and hence by Theorem XI.1, $\mathrm{Im}\,(L^t - \tilde{T}) = \tilde{X}$. If now $\ker (L - B) \neq \{0\}$, then, if $\mathrm{Im}\,(T - B) \subset (\ker (L^t - B^t))^{\perp}$, $\mathrm{Im}\,(T - B) \subset \mathrm{Im}\,(L -$ by Lemma XI.4. (b) and hence $\mathrm{Im}\,(L - T) = \mathrm{Im}\,(L - B) = (\ker (L^t - B^t))^{\perp}$ by the same Lemma and Theorem XI.1 and the reasoning is the same if $\mathrm{Im}\,(\tilde{T} - B^t) \subset (\ker (L - B))^{\perp}$.

Remark. The results above can be formulated and similarly proved for k-ball contractions, and combinations "k-ball-contractions" and "k-set contractions" for $T,\tilde{T}$ are also allowed.

7. We shall now obtain a result which implies the existence of at least three solutions.

Lemma XI.5. *Let* $L : \mathrm{dom}\, L \subset X \to Z$ *be linear, closed and Fredholm of index zero,* $k \in [0,l(L)[$, *and* $N : \bar{\Omega} \subset X \to Z$ *a* k*-set contraction which is Frechet-differentiable at* $\bar{x} \in \mathrm{dom}\, L \cap \Omega$ *with* $N'(\bar{x})$ *such that* $\ker (L - N'(\bar{x})) = \{0\}$.
Then, if $L\bar{x} = N\bar{x}$,

(1) $\bar{x}$ *is a isolated coincidence point of* L *and* N.

(2) *For each* $\varepsilon > 0$ *such that* $\bar{B}(\bar{x}.\varepsilon) \subset \Omega$ *and* $(L - N)^{-1}(0) \cap \bar{B}(\bar{x},\varepsilon) = \{\bar{x}\}$ *one has*

$$d\,[(L,N), B(\bar{x},\varepsilon)] = d\,[(L, N'(\bar{x})], B(0,\varepsilon)] = \pm 1$$

Proof. By assumption $L - N'(\bar{x})$ is one-to-one and hence because of $N'(\bar{x})$ is also a k-set contraction, $L - N'(\bar{x})$ is a closed Fredholm mapping of index zero (Proposition XI.3) which is therefore onto with $[L - N'(\bar{x})]^{-1}$ continuous. Therefore, for some $m > 0$, one has

$$|(L - N'(\bar{x}))(x - \bar{x})| \geq m|x - \bar{x}|$$

for all $x \in \operatorname{dom} L$. Now there exists $\varepsilon > 0$ such that $\bar{B}(\bar{x},\varepsilon) \subset \Omega$ and

$$|R(x,\bar{x})| = |Nx - N\bar{x} - N'(\bar{x})(x - \bar{x})| \leq \frac{m}{2}|x - \bar{x}|$$

for $x \in \bar{B}(\bar{x},\varepsilon)$. This implies that for $x \in \bar{B}(\bar{x},\varepsilon)$,

$$|Lx - Nx| \geq \frac{m}{2}|x - \bar{x}|$$

and $\bar{x}$ is isolated. Now, if

$$\tilde{N}(x,\lambda) = \lambda Nx + (1 - \lambda)(N'(\bar{x})(x - \bar{x}) + L\bar{x}),$$

we have if $x \in \partial B(\bar{x},\varepsilon)$

$$|Lx - \tilde{N}(x,\lambda)| = |Lx - Nx + (1 - \lambda)(Nx - N'(\bar{x})(x - \bar{x}) - L\bar{x}|$$

$$= |Lx - Nx + (1 - \lambda)(Nx - N\bar{x} - N'(\bar{x})(x - \bar{x}))| \geq \frac{m}{2}|x - \bar{x}| = \frac{m}{2}\varepsilon > 0$$

and hence, $\tilde{N}$ being clearly a k-set contraction on $\bar{B}(\bar{x},\varepsilon) \times [0,1]$,

$$d[(L,N), B(\bar{x},\varepsilon)] = d[(L, \tilde{N}(.,1)), B(\bar{x},\varepsilon)] = d[(L, \tilde{N}(.,0)), B(\bar{x},\varepsilon)] =$$

$$= d[(L, N'(\bar{x})(. - \bar{x}) + L\bar{x}), B(\bar{x},\varepsilon)]$$

$$= d[(L, N'(\bar{x}), B(0,\varepsilon)] = \pm 1.$$

by usual properties of the degree of linear k-set contractive perturbations of identity.

Theorem XI.3. *Let* $L : \operatorname{dom} L \subset X \to Z$ *be linear, closed and Fredholm of index zero, let* $k \in [0,l(L)[$, $T : X \to Z$ *be asymptotically linear with an asymptotic derivative* B *such that* $L - B$ *is one-to-one. Let* $x_1, x_2 \in \operatorname{dom} L$, $x_1 \neq x_2$ *be such that* T *has Frechet derivatives at* x_1 *and* x_2 *and such that*

$$Lx_i = Tx_i + y \quad (i = 1,2)$$

for some $y \in Z$. *If* $\ker(L - T'(x_i)) = \{0\}$ $(i = 1,2)$, *then there exists*

$x \in \text{dom } L$ *with* $x \neq x_i$ $(i = 1,2)$ *such that*

$$Lx = Tx + y.$$

Proof. Without loosing generality we can assume that $y = 0$. It follows from the proof of Lemma XI.3 that some $\rho_0 > 0$ exists such that, for all $\rho \geq \rho_0$,

$$d[(L,T), B(0,\rho)] = d[(L,B), B(0,\rho)] = \pm 1$$

Let $\tilde{\rho} \geq \rho_0$ such that $x_i \in B(0,\tilde{\rho})$, $i = 1,2$. By assumptions, and Lemma XI.5, x_i are isolated coincidence points of L and T and hence for $\varepsilon_1 > 0$, $\varepsilon_2 > 0$ sufficiently small,

$$\pm 1 = d[(L,T), B(0,\tilde{\rho})] = d[(L,T), B(x_1,\varepsilon_1)] + d[(L,T), B(x_2,\varepsilon_2)]$$
$$+ d[(L,T), B(0,\tilde{\rho}) \setminus \bigcup_{i=1}^{2} \overline{B}(x_i,\varepsilon_i)]$$
$$= d[(L,T'(x_1)),B(0,\varepsilon_1)] + d[(L,T'(x_2)),B(0,\varepsilon_2)]$$
$$+ d[(L,T), B(0,\tilde{\rho}) \setminus \bigcup_{i=1}^{2} \overline{B}(x_i,\varepsilon_i)]$$
$$= d[(L,T), B(0,\tilde{\rho}) \setminus \bigcup_{i=1}^{2} \overline{B}(x_i,\varepsilon_i)] \qquad (\text{mod } 2)$$

Hence $d[(L,T), B(0,\tilde{\rho}) \setminus \bigcup_{i=1}^{2} \overline{B}(x_i,\varepsilon_i)] \neq 0$ and the result follows.

Remark. An analogous result holds for k-ball contractions.

8. We shall now develope the relations of the obtained coincidence degree with the classical *alternative problems*. Let us assume that the L-k-set contractive character of N is replaced by the condition

(C) *There exists* $k \in [0,1[$ *such that, for each* $x,y \in \overline{\Omega}$,

$$|K_{P,Q}(Nx - Ny)| \leq k\,|x - y|.$$

Let us define $H : \overline{\Omega} \to X$ by

$$H = I - K_{P,Q} N .$$

Proposition XI.5. *H is a homeomorphism of* Ω *onto an open set* $H(\Omega)$ *of* X *which maps* $\overline{\Omega}$ *homeomorphically onto* $\overline{H(\Omega)}$. *In particular,* $H(\partial\Omega) = \partial H(\Omega)$. *Also,* H *and* H^{-1} *map bounded sets onto bounded sets.*

Proof. Let us write for simplicity $K_{P,Q} N = G$. Then

$$|Gx - Gy| \leq k |x - y|$$

for all $x, y \in \overline{\Omega}$. If $u, v \in \mathrm{cl}\,\Omega$ and $w = Hu$, $z = Hv$, one has

$$|w - z| = |u - v + Gu - Gv| \geq (1 - k)|u - v|$$

which shows that H is bijective form $\overline{\Omega}$ onto $H(\overline{\Omega})$ and

$$|H^{-1} w - H^{-1} z| \leq (1 - k)^{-1} |w - z| \qquad \text{(XI.5)}$$

for all $w, z \in H(\overline{\Omega})$. Thus H is an homeomorphism from $\overline{\Omega}$ was $H(\overline{\Omega})$, with H and H^{-1} lipschitzians.

Let us show now that $H(\Omega)$ is open and let $v_o \in H(\Omega)$. Thus there exists an unique $u_o \in \Omega$ such that

$$v_o = H(u_o) .$$

and there exists $d > 0$ such that the closed ball $B_d(u_o) \subset \Omega$. We shall show that there exists $d_1 > 0$ such that each $v \in B_{d_1}(v_o)$ is in $H(B_d(u_o))$ and hence in $H(\Omega)$. So let $v \in B_d(v_o)$ and consider the equation

$$H u = v$$

or, writing,

$$u = u_o + \tilde{u}, \; |\tilde{u}| \leq d$$

$$v = v_o + \tilde{v}, \; |\tilde{v}| \leq d_1,$$

$$u_o + \tilde{u} - G(u_o + \tilde{u}) = v_o + \tilde{v} ,$$

i.e.

$$\begin{aligned} \tilde{u} &= G(u_o + \tilde{u}) - u_o + H(u_o) + \tilde{v} \\ &= G(u_o + \tilde{u}) - G(u_o) + \tilde{v} = S(\tilde{u}). \end{aligned}$$

Also, if $\tilde{u}, \overline{u} \in B_d(u_o)$,

$$\begin{aligned} |S(\tilde{u}) - S(\overline{u})| &= |G(u_o + \tilde{u}) - G(u_o + \overline{u})| \\ &\leq k |\tilde{u} - \overline{u}| \end{aligned}$$

and

$$|S(\tilde{u})| \leq |S(\tilde{u}) - S(0)| + |S(0)| \leq k|\tilde{u}| + |\tilde{v}| \leq kd + d_1$$

$$\leq d$$

if we choose

$$d_1 \leq (1 - k)d \, .$$

Then, by the Banach fixed point principle, S has a fixed point for each z in $B_{d_1}(0)$ i.e. $H(B_d(u_o))$ contains $B_{(1-k)d}(H(u_o))$ and $H(\Omega)$ is open.

Let us show now that $H(\overline{\Omega})$ is closed in X. Let $\{w_j\}$ a sequence in $H(\overline{\Omega})$ which converges to $w \in X$. Then

$$w_j = H(u_j) \ , \ u_j \in \overline{\Omega}$$

and, using (XI.5),

$$|u_j - u_l| = |H^{-1}(w_j) - H^{-1}(w_l)| \leq (1 - k)^{-1} |w_j - w_l|$$

which implies that $\{u_j\}$ is a Cauchy sequence and hence converges to $u \in \overline{\Omega}$. Thus, $w = H(u) \in H(\overline{\Omega})$ and $H(\overline{\Omega})$ is closed. Therefore,

$$H(\overline{\Omega}) \supset \overline{H(\Omega)}$$

but one also has

$$H(\overline{\Omega}) \subset \overline{H(\Omega)}$$

because if $w \in H(\overline{\Omega})$, $w = H(u)$ with $u = \lim_{j \to \infty} u_j$ and $u_j \in \Omega$, and hence

$$w = H(u) = \lim_{j \to \infty} H(u_j) = \lim_{j \to \infty} w_j$$

with $w_j = H(u_j) \in H(\Omega)$, which implies that $w \in \overline{H(\Omega)}$. Now,

$$H(\partial\Omega) = H(\overline{\Omega}) \setminus H(\Omega) = \overline{H(\Omega)} \setminus H(\Omega) = \partial H(\Omega) \, .$$

The last assertion follows directly from (XI.5) and from the Lipschitzian character of H.

Proposition XI.6. H *and* H^{-1} *preserve the fibres of* P, *i.e.*

$$PHx = Px \ , \ x \in \overline{\Omega}, \ P H^{-1} y = Py, \ y \in \overline{H(\Omega)} \qquad \text{(XI.6)}$$

and, if $\Sigma = \Omega$ *or* $\overline{\Omega}$,

$$PH(\Sigma) = P(\Sigma), \ H(\Sigma) \cap \text{Im } P \subset P(\Sigma). \qquad \text{(XI.7)}$$

Proof. The first relation in (XI.6) is trivial and the second one follows immediately from it and from the fact that $H(\overline{\Omega}) = \overline{H(\Omega)}$. The first relation (XI.7) is a direct consequence of (XI.6) and if $x \in H(\Sigma) \cap \text{Im } P$, $x = Px$ and $x = H(y)$, $y \in \Sigma$, and hence $x = Px = Py \in P(\Sigma)$.

Theorem IX.4. *If* $L : \text{dom } L \subset X \to Z$ *is a Fredholm mapping of index zero and* $N : \overline{\Omega} \subset X \to Z$ *satisfies condition* (C), *then*

$$d \ [(L,N),\Omega] = d_B \ [- \Lambda\Pi N H^{-1} | \ker L, \ H(\Omega) \cap \ker L, \ 0]$$

where the right hand member is a Brouwer degree.

Proof. Let $C : \overline{\Omega} \to X$, $H : \overline{\Omega} \to X$ and $G : \overline{\Omega} \to X$ be defined by

$$C = \Lambda\Pi N + P \ , \ G = K_{P,Q} N \ , \ H = I - G \ ,$$

and, for each $\lambda \in [0,1]$, let $H_\lambda : \overline{\Omega} \to X$ be defined by

$$H_\lambda = I - \lambda G \ .$$

Clearly H and each H_λ satisfies the properties given in Proposition XI.5 and property (C). Consider, for each $\lambda \in [0,1]$, the mapping

$$(I - CH^{-1})H_\lambda : H_\lambda^{-1} [H(\overline{\Omega})] \to X \ .$$

Explicitely,

$$(I - CH^{-1}) \ H_\lambda = I - \lambda G - CH^{-1} \ H_\lambda$$

and

$$(I - CH^{-1}) \ H_0 = I - CH^{-1}$$
$$(I - CH^{-1}) \ H_1 = H - C = I - C - K = I - M_\Lambda \ .$$

Then if we define $F : \bigcup_{\lambda \in [0,1]} [H_\lambda^{-1} [H(\overline{\Omega})] \times \{\lambda\}] \to X$ by

$$F(x,\lambda) = \lambda Gx + CH^{-1} (I - \lambda G)x$$

and if we write

$$\Delta = \bigcup_{\lambda \in [0,1]} [H_\lambda^{-1} [H(\Omega)] \times \{\lambda\}],$$

Δ is an open subset of $X \times [0,1]$, $F : \overline{\Delta} \to X$ and, for each bounded $A \subset X$,

$$\alpha \{F [\Omega \cap (A \times [0,1])]\}$$

$$\leq \alpha \{ \bigcup_{\lambda \in [0,1]} \lambda G [H_\lambda^{-1} H(\Omega) \cap A]\}$$

$$+ \alpha \{ \bigcup_{\lambda \in [0,1]} CH^{-1} H_\lambda [H^{-1} H(\overline{\Omega}) \cap A]\}$$

$$\leq \alpha [\text{co} \{G [\bigcup_{\lambda \in [0,1]} H^{-1} H(\Omega) \cap A] \cup \{0\}]$$

$$+ \alpha [\bigcup_{\lambda \in [0,1]} C(\overline{\Omega}) \cap C H^{-1} H_\lambda(A)]$$

$$\leq \alpha \; G [\bigcup_{\lambda \in [0,1]} H_\lambda^{-1} H(\Omega) \cap A] + \alpha [C(\overline{\Omega}) \cap \bigcup_{\lambda \in [0,1]} CH^{-1} H_\lambda(A)]$$

$$\leq k \; \alpha(A)$$

using the fact that G is a k-set contraction and C is a o-set contraction. Thus, the conditions of the theorem of invariance with respect to homotopy for k-set contractions will be satisfied if we verify that

$$x \neq F(x,\lambda)$$

for each $(x,\lambda) \in \partial\Delta$, the boundary of Δ in $X \times [0,1]$. But, if there is an $x \in \partial\Delta$ such that

$$x = F(x,\lambda)$$

then, using Proposition XI.5,

$$H_\lambda x = C H^{-1} H_\lambda x$$

for some $\lambda \in [0,1]$ and some $x \in \partial [H_\lambda^{-1} H(\Omega)] = H_\lambda^{-1} H(\partial\Omega)$ and hence

$$y_\lambda = H_\lambda x \in H(\partial\Omega)$$

is such that

$$y_\lambda = C H^{-1} y_\lambda .$$

Therefore,

$$z_\lambda = H^{-1} y_\lambda \in \partial\Omega$$

is such that

$$Hz_\lambda = Cz_\lambda$$

i.e.

$$z_\lambda = M_\Lambda z_\lambda ,$$

a contradiction with the assumptions. Therefore,

$$d[(L,N),\Omega] = d[I - F(.,1),\Omega,0] =$$

$$d[I - F(.,0), H(\Omega), 0] = d[I - CH^{-1}, H(\Omega), 0], \qquad (XI.8)$$

and CH^{-1} being compact, the last member of (XI.8) is a Leray-Schauder degree. But, using Proposition XI.6,

$$CH^{-1} = \Lambda\Pi NH^{-1} + PH^{-1} = \Lambda\Pi NH^{-1} + P$$

and hence,

$$d[I - CH^{-1}, H(\Omega), 0] = d[I - P - \Lambda\Pi NH^{-1}, H(\Omega), 0]$$

$$= d_B[(I - P - \Lambda\Pi NH^{-1})| \ker L, H(\Omega) \cap \ker L, 0]$$

$$= d_B[-\Lambda\Pi NH^{-1} | \ker L, H(\Omega) \cap \ker L, 0].$$

Now Theorem XI.4 can be improved if one makes a supplementary assumption which is classical in alternative methods.

<u>Theorem XI.5</u>. *If conditions of Theorem* XI.4 *hold and if, for each* $a \in P(\bar{\Omega})$, *the mapping*

$$T_a : x \to a + K_{P,Q} Nx$$

maps $\bar{\Omega}$ *into itself and is such that if* $a \in P(\Omega)$, $T_a(\Omega) \subset \Omega$, *then*

$$d\,[\,(L,N),\Omega\,] = d_B\,[\,-\Lambda\Pi NR\,|\,\ker L,\ P(\Omega), 0\,]$$

where R(a) *is the (unique) solution of the equation*

$$y = a + K_{P,Q}\,Ny\ .$$

Proof. By the assumptions and Banach fixed point theorem, R(a) exists and is unique for each $a \in P(\overline{\Omega})$, $R : P(\overline{\Omega}) \to X$ is continuous and $R(a) \subset \Omega$ if $a \in P(\Omega)$. which implies that $H(\Omega) \supset P(\Omega)$ and therefore, using (XI.7), $H(\Omega) \cap \ker L = P(\Omega)$. The result follows then from Theorem XI.4.

If we note that $\Lambda\Pi N$ can be written JQN with $J : \text{Im } Q \to \ker L$ some isomorphism, we note that $-\Lambda\Pi R$ is nothing but the mapping defining the *bifurcation equations* of the corresponding alternative method. Hence Theorem XI.5 explicitely relates the coincidence degree of L and N in Ω with the Brouwer degree of the bifurcation mapping at zero in $P(\Omega)$.

9. Bibliographical notes about Chapter XI.

The measure of noncompactness α is due to Kuratowski (*Fundam. Math.* 15 (1930) 301 - 309) and the concept of k-set contraction to Darbo (*Rend. Sem. Mat. Univ. Padova* 24 (1955) 84 - 92). Degree theory has been extended to k-set contractive perturbations of identity in Banach spaces by Nussbaum (*Ann. Mat. Pura Appl.* (4) 89 (1971) 217 - 258, *J. Math. Anal. Appl.* 37 (1972) 741 - 766). Coincidence degree has been extended to the situation described in section 3 by Hetzer (*Ann. Soc. Sci. Bruxelles* 89 (1975) 497 - 508). The measure of non-compactness χ_M has been introduced by Gol'denshtein, Gokhberg and Markus (*Uchen. Zap. Kishinev univ.* 29 (1957) 29 - 36) and degree theory for k-ball contractive perturbations of identity in Banach spaces is due to Vainikko and Sadovskii (*Probl. Matem. Analiza Slozhn. Sist.* n°2, Voronezh (1968) 84 - 88) and Borisovich and Sapronov (*Soviet Math. Dokl. 9 (1968)* 1304 - 1308). See Sadovskii (*Russian Math. Surveys* 27 (1972) 85 - 156) and Danes (Proc. *"Theory of Nonlinear Operators"*, Akademic-Verlag, Berlin, 1974, 15 - 56) for references about set contractions. Proposition XI.1, XI.2 and XI.3 are due to Hetzer (*op. cit.*) as well as Definition XI.3, and extend previous results of Lebow and Schechter (*J. Functional Analysis* 7 (1971) 1 - 26). The extension, to k-set contractive perturbations N of closed Fredholm linear mapping of index zero in Banach spaces, when $k \in [\,0, l(L)\,[$, of Corollary IV.1 and Theorem VII.1 can also be found in Hetzer (op. cit.) and they are applied in Hetzer (*Comment. Mathem. Universit. Carolinae* 16 (1975) 121 - 138) to the periodic solutions of neutral functional differential equations. See also Sadovskii (Soviet Math. Dokl. 12 (1971) 1543 - 1547) for the statement of analogous results for neutral equations. Theorem VII.4 has also been extended to the case of a k-set ontractive perturbation by Hetzer (On the existence of weak solutions for some uasilinear elliptic variational boundary value problems at resonance, to appear) nd used in a manner similar to the one used in Theorem VIII.2 to extend to ome quasilinear elliptic equations results of de Figueiredo concerning the emilinear case ("Partial Differential Equations and related topics", Springer ecture Notes n° 446 (1975), Springer-Verlag, Berlin). The concept of asympoti-ally linear mapping is due to Krasnosel'skii (*Uspekhi Mat. Nauk* 9 (1954) 57 - 114) nd Proposition XI.4 is due to Amann (*J. Funct. Anal.* 14 (1973) 162 - 171). emma XI.3 and Theorems XI.1 and XI.2 are due to Hetzer and Stallbohm (Eine xistenzaufssage für asymptotisch lineare Störungen eines Fredholmoperators it Index 0, to appear) and generalize previous results of Kačurovskii (*Soviet ath. Dokl.* 11 (1970) 751 - 754; 12 (1971) 168 - 172) and Petryshyn (*J. Differential quations* 17 (1975) 82 - 95). Lemma XI.5 and Theorem XI.3 are due to Hetzer and tallbohm (*op. cit.*) and generalize results of Amann (*Indiana Univ. Math. J.* 1 (1972) 925 - 935) and Petryshyn (*op. cit.*). The results of section 8 are due o Mawhin (Proc. Intern. Sympos. Dynamical Systems, Providence, 1974, to appear) xcept for the concept of permissible homeomorphism and Proposition XI.5 which ome from Browder (Proc. Symp. Pure Math., vol. 18, II, AMS, Providence, 1976). he proof of Theorem IX.4 given in Mawhin's original paper is different and makes se of various generalized degrees of Browder (op. cit. and *Studia Math.* 31 (1968) 89 - 204)), Browder and Nussbaum (*Bull. Am. Math. Soc.* 74 (1968) 671 - 676) and ussbaum (*J. Math. Anal. Appl.* 37 (1972) 741 - 766). Theorem XI.5 contains a pecial cases results of O'Neil and Thomas (*Trans. Amer. Math. Soc.* 167 (1972) 33 - 345)) and Thomas (*Duke Math. J.* 40 (1973) 233 - 240; *Scripta Math.* to appear) elating Browder - Nussbaum - Petryshyn degrees to Cronin's concept of multiplicity *Trans. Amer. Math. Soc.* 69 (1950) 208 - 231; 76 (1954) 207 - 222; *Amer. J. Math.* 3 (1951) 763 - 772) which is nothing but the Brouwer degree of the bifurcation apping of an alternative problem with $L = I - A$, A compact. See also Williams *Michigan Math. J.* 15 (1968) 441 - 448) for a corresponding result in the line of esari's alternative method. Let us also mention that the results of Chapter X bout bifurcation theory have been partially extended, with similar proofs, by etzer and Stallbohm ("Coincidence degree and Rabinowitz bifurcation theorem", o appear), to the case of k-set contractive perturbations of closed linear Fredholm appings of index zero with $k \in [0,l(L)[$. One shall find in particular in this aper the extension of Lemma X.3, Theorem X.7 and, under the assumption (H_6') of hapter X, the extension of the concept of multiplicity for a characteristic value sing the result of Theorem X.4 as definition.

Under this assumption (H_6'), one also finds in Hetzer and Stallbohm's paper the generalization of Theorem X.5 and Theorem X.8. Lastly the global bifurcation results are also considered and a theorem is given which extends previous ones of Rabinowitz (*J. Functional Analysis* 7 (1971) 487 - 513) and Stuart (*Proc. London Math. Soc.* (3) 27 (1973) 531 - 550).

XII. NONLINEAR PERTURBATIONS OF FREDHOLM MAPPINGS OF NONZERO INDEX

1. All the results exposed up to now concern nonlinear perturbations of inear Fredholm mappings with zero index. We shall now try to drop this ssumption. Let us take the notations of section A of chapter III and let s note that a linear one-to-one mapping

$$\Lambda : \text{coker } L \to \ker L$$

xists if and only if $\dim \ker L \geq \dim \text{coker } L$, i.e. if and only if

$$\text{Ind } L \geq 0.$$

nder this assumption, Proposition III.1 holds and implies that

$$Lx = Nx$$

f and only if

$$(I - P)x = (\Lambda\Pi + K_{P,Q})Nx$$

or any linear one-to-one $\Lambda : \text{coker } L \to \ker L$.

Assume now that X, Z are real normed spaces, $L : \text{dom } L \subset X \to Z$ is a redholm mapping with

$$\text{Ind } L \geq 0,$$

$\subset X$ an open bounded set and $N : \bar{\Omega} \to Z$ is L-compact. Then the following negative" result holds.

<u>Proposition XII.1</u>. *If* $\text{Ind } L > 0$ *and*

$$0 \notin (L - N)(\text{dom } L \cap \partial\Omega),$$

hen for each linear one-to-one $\Lambda : \text{coker } L \to \ker L$, *one has*

$$d\,[I - P - (\Lambda\Pi + K_{P,Q})N, \Omega, 0] = 0.$$

<u>Proof</u>. Because of $\dim \ker L > \dim \text{coker } L$, $\text{Im } \Lambda$ is a proper subspace of er L and therefore the range of the mapping

$$I - M = I - P - K_{P,Q}N - \Lambda\Pi N$$

s necessarily contained in the proper subspace of X given by

$$X' = \ker P \oplus \text{Im } \Lambda .$$

the properties of Leray-Schauder degree there exists a neighbourhood V of e origin such that

$$d[I - M,\Omega,0] = d[I - M,\Omega,y]$$

for all $y \in V$. If we take y in the nonvoid set $V \cap \complement_X X'$, then y does not belong to the range of I - M and consequently

$$d[I - M,\Omega,y] = 0$$

which achieves the proof.

This "negative" result can be used as an existence tool.

Theorem XII.1. *If* $\mathrm{Ind}\, L > 0$ *and if* $N : X \to Z$ *is* L-*compact on bounded sets of* X *and odd, then for each open, bounded symmetric neighbourhood* $\mathcal{O}$ *of* 0 *in* X *there exists* $x \in \mathrm{dom}\, L \cap \partial\mathcal{O}$ *such that*

$$Lx - Nx = 0 = L(-x) - N(-x).$$

Proof. If not, there will exist an open, bounded symmetric neighbourhood $\mathcal{O}$ of 0 such that

$$0 \notin (L - N)(\mathrm{dom}\, L \cap \partial\mathcal{O})$$

and hence, using Borsuk theorem for the degree of odd mappings and the fact that

$$0 \notin (I - M)(\partial\, \mathcal{O})$$

we obtain

$$d[I - M,\mathcal{O},0] = 1 \pmod 2,$$

a contradiction with Proposition XII.1.

Corollary XII.1. *Under assumptions of Theorem* XII.1, *the set of zeros of* L - N *possesses a symmetric unbounded component* $\mathcal{C}$ *containing* x = 0.

Proof. Note that the components of zeros of L - N occur in antipodal pairs because of the oddness of L - N. Suppose the statement of Corollary XII.1 is not rue. Then x = 0 belongs to a symmetric bounded component $\mathcal{C}$ of zeros of L - N. Let $\mathcal{O}$ be a δ-neighbourhood of $\mathcal{C}$ and let S be the set of zeros of L - N. As S is also in the set of fixed points of the compact mapping M we note that $S \cap \overline{\mathcal{O}}$ is a compact metric space under the topology induced by X. Let $A = \partial\mathcal{O} \cap S$ and $B = \mathcal{C}$. By a known result there exist disjoint compact sets $\hat{A}, \hat{B} \subset \overline{\mathcal{O}}$ such that $\hat{A} \supset A$, $\hat{B} \supset B$ and

$$S \cap \overline{\mathcal{O}} = \hat{A} \cup \hat{B}$$

Let A^{ε} be an ε-neighbourhood of $\hat{B}$ where ε is less than the distance from $\hat{A}$ to $\hat{B}$.

Then $\mathcal{B} \subset A$ and $\partial A \cap S = \emptyset$. It is clear that A is a symmetric neighbourhood of 0 but ∂A contains no zero of L - N, a contradiction with Theorem XII.1.

One can sharpen Theorem XII.1 by introducing the concept of genus of a set. If A, B are metric spaces, let C(A,B) be the space of continuous maps from A to B. Let E be a Banach space and $\Sigma(E) \equiv \Sigma$ be the set of closed (in E) subsets A of $E \setminus \{0\}$ which are symmetric with respect to the origin.

Definition XII.1. For $A \in \Sigma$, the *genus* $\gamma(A)$ of A is the smallest integer n such that there exists an odd continuous map $\phi \in C(A, R^n \setminus \{0\})$.

By definition $\gamma(\emptyset) = 0$ and if there is no such n, $\gamma(A) = \infty$. It is easy to see that if $\gamma(A) = 1$, A is not connected. We shall give without proof some of the properties of the genus.

Proposition XII.2. *Let* $A, B \in \Sigma(E)$.

1° *If there exists an odd* $\phi \in C(A,B)$, *then* $\gamma(A) \leq \gamma(B)$

2° *If* $A \subset B$, *then* $\gamma(A) \leq \gamma(B)$

3° $\gamma(A \cup B) \leq \gamma(A) + \gamma(B)$

4° *If* $\gamma(A) < \infty$, *then* $\gamma(\overline{A \setminus B}) \geq \gamma(A) - \gamma(B)$

5° *If* A *is compact, then* $\gamma(A) < \infty$

6° *If* A *is compact there exists an uniform neighbourhood*

$N_\gamma(A) = \{u \in E : |u - A| \leq \gamma\}$ *of* A *such that*

$$\gamma(N_\delta(A)) = \gamma(A).$$

7° *If* A *is homeomorphic by an odd homeomorphism to the unit sphere* S^{n-1} *in* R^n, *then* $\gamma(A) = n$.

We can now prove the following

Theorem XII.2. *Suppose* $\mathcal{O}$ *is a symmetric bounded open neighbourhood of* 0 *in* R^q *and* $\phi \in C(\partial\mathcal{O}, F)$ *where* ϕ *is odd and* F *is a proper vector subspace of* R^q *which is isomorphic to* R^p. *If* $G = \{x \in \partial\mathcal{O} : \phi(x) = 0\}$, *then* $\gamma(Z) \geq$ - p.

Proof. By 6° of Proposition XII.2 there exists a neighbourhood of G, $N_j(G)$ such that $\gamma(N_j(G)) = \gamma(G)$. Since ϕ is continuous on $\partial\mathcal{O}$ and G is compact it can further be assumed that there exists $\varepsilon_0 > 0$ such that

$$N_\delta(G) \supset G_\varepsilon = \{x \in \partial\mathcal{O} : |\phi(x)| \leq \varepsilon\}$$

for all $\varepsilon \leq \varepsilon_0$.

By 2° of Proposition XII.2, $\gamma(G_\varepsilon) = \gamma(G)$. Let $Y_\eta = \{x \in \partial\mathcal{O} : |\phi(x)| \geq \eta\}$ where $\eta > 0$ and let $\pi(x) = x/|x|$ for $x \in R^q \setminus \{0\}$. Then $\pi \circ \phi$ is odd and belongs to $C(Y_\eta, S_F)$ with S_F the unit sphere in F. Since F is isomorphic to R^p, then by 1° and 7° of Proposition XII.2, $\gamma(Y_\eta) \leq \gamma(S_F) = p$. Hence by 4° of the same Proposition,

$$\gamma\,\overline{(\partial\mathcal{O} \setminus Y_\eta)} \geq \gamma(\partial\mathcal{O}) - \gamma(Y_\eta) \geq \gamma(\partial\mathcal{O}) - p.$$

Using for example Tietze theorem we can extend ϕ to an element in $C(\overline{\mathcal{O}}, F)$. Then $\gamma(\partial\mathcal{O}) = q$ because by a reasoning analogous to the one used in Theorem XII.1 any continuous mapping ψ which is odd on $\partial\mathcal{O}$ and such that $\psi(0)$ is a proper subspace of R^q is such that $0 \in \psi(\partial\mathcal{O})$ and on the other hand $I \in C(\partial\mathcal{O}, R^q \setminus \{0\})$. Thus $\gamma\overline{(\partial\mathcal{O} \setminus Y_\eta)} \geq q - p$. But $\overline{\partial\mathcal{O} - Y_\eta} = G_\eta$ so if we choose $\eta = \varepsilon \leq \varepsilon_0$, we get $\gamma(G_\varepsilon) = \gamma(G) \geq q - p$.

Theorem XII.3. *Under the assumptions of Theorem* XII.1, *if* $\mathcal{O}$ *is a symmetric bounded neighbourhood of* 0 *and if*

$$G = \{x \in \text{dom } L \cap \partial\mathcal{O} : Lx = Nx\}$$

then $\gamma(G) \geq \text{ind } L$.

Proof. By the assumptions G is the set of fixed points of M on $\partial\mathcal{O}$ and is then a compact subset of X symmetric with respect to the origin. Thus by Proposition XII.2.5°, $\gamma(G) < \infty$ and by 6° G has an uniform neighbourhood $N_\gamma(G)$ such that $\gamma(N_\gamma(G)) = \gamma(G)$. Now there exists $\eta_0 > 0$ such that $N_\gamma(G) \supset G\eta = \{x \in \partial\mathcal{O} : |x - Mx| \leq \eta\}$ for all $\eta \leq \eta_0$; because if not there is a sequence $(u_n) \subset \partial\mathcal{O} \setminus N_\delta(G)$ such that $|x_n - Mx_n| \leq \frac{1}{n}$. But (Mx_n) has a convergent subsequence and hence the same is true for (x_n). Let $x_{n_i} \to x \in \overline{\partial\mathcal{O} \setminus N_\gamma(G)}$. Then $x = Mx$, which is impossible because $x \notin G$. Thus $N_\delta(G) \supset G_\eta$ and by Proposition XII.2, 2°, $\gamma(G_\eta) = \gamma(G)$. Now $K_{P,Q}N$ can be approximated by a finite dimensional odd map with any accuracy by using an easy modification of the standard Leray-Schauder argument. Let S_ε be such a map for which $|S_\varepsilon x - K_{P,Q}Nx| \leq \varepsilon$ for all $x \in \partial\mathcal{O}$, and let

$$M_\varepsilon = P + \Lambda\Pi N + S_\varepsilon .$$

and X_ε finite-dimensional such that $\text{Im } S_\varepsilon \subset X_\varepsilon \subset \ker P$. Let $m = \dim \ker L$ and $n = \dim \text{coker } L$. Then

$$I - M_\varepsilon = I - P - S_\varepsilon - \Lambda\Pi N$$

is an odd continuous mapping from $\partial O \cap (X_\varepsilon \oplus \ker L)$ into $X_\varepsilon \oplus \operatorname{Im} \Lambda$. By Theorem XII.2,

$$\gamma \{\{x \in \partial O \cap (X_\varepsilon \oplus \ker L) : x = M_\varepsilon x\} \geqslant \dim \ker L - \dim \operatorname{coker} L$$

$$= \operatorname{ind} L$$

and hence by Proposition XII.1 - 2°,

$$\gamma(\{x \in \partial O : |x - M_\varepsilon x| \leqslant \varepsilon\}) \geqslant \operatorname{ind} L .$$

Now if $x \in \partial O$ and $x = Mx$, then

$$|x - M_\varepsilon x| \leqslant |K_{P,Q} Nx - S_\varepsilon x| \leqslant \varepsilon$$

and if $x \in \partial O$ and $|x - M_\varepsilon x| \leqslant \varepsilon$, then

$$|x - Mx| \leqslant |x - M_\varepsilon x| + |K_{P,Q} Nx - S_\varepsilon x| \leqslant 2\varepsilon .$$

Therefore

$$\gamma(\{x \in \partial O : |x - Mx| \leqslant 2\varepsilon\}) \geqslant \gamma \{x \in \partial O : |x - Mx| \leqslant \varepsilon\} \geqslant \gamma(G)$$

and choosing $2\varepsilon \leqslant \eta_o$, we obtain

$$\gamma(G) = \gamma(G_{2\varepsilon}) \geqslant \operatorname{ind} L .$$

As a direct application let us consider the elliptic boundary value problem

$$Lu \equiv \sum_{|\sigma| \leqslant 2m} a_\sigma(x) D^\sigma u = g(x,u), \; x \in \Omega$$

$$B_i u \equiv \sum_{|\sigma| \leqslant m_i} b_{i\sigma}(x) D^\sigma u = 0 \quad , x \in \partial\Omega$$

where $1 \leqslant i \leqslant m$, $m_i < 2m$, $x = (x_1,\dots,x_n) \in \Omega$ which is a smooth bounded domain in R^n, L is uniformly elliptic with smooth coefficients, g is continuously differentiable on $\overline{\Omega} \times R$, the B_i have smooth coefficients and the usual multi-index notation is being employed. Also the boundary conditions are assumed to be complementing which implies that L is a Fredholm mapping with dense domain in the Banach space X of functions which are Hölder continuous with coefficient $\alpha < 1$ and which satisfy the boundary conditions. We shall assume that this Fredholm mapping has a strictly positive index. It is also known that the right inverses of L are compact and then if $L = L|\text{dom } L \cap X$ and $N : X \to X$ is defined by

$$N : X \to X \; , \; u \to g(.,u) \; ,$$

then N is L-compact on bounded sets. One immediately gets from Theorems XII.1 and XII.3 the following.

Theorem XII.3'. *Under assumptions above, if* g *is odd in* u *then the elliptic problem above has a symmetric unbounded component* $\mathcal{C}$ *of solutions containing the origin and if*

$$G = \{x \in \text{dom } L \cap \mathcal{O} : Lu = Nu\}$$

for $\mathcal{O}$ *a symmetric bounded neighbourhood of* 0, *then* $\gamma(G) \geq \text{Ind } L$.

2. One shall now overcome the difficulty created by Proposition XII.1 by modifying the operator M_Λ related to $L - N$ in such a way that the degree is no more necessarily equal to zero. We still assume that Ind $L \geq 0$ so that one-to-one linear mappings Λ : coker $L \to$ ker L do exist. We have now the following.

Proposition XII.3. *Equation*

$$Lx = Nx \qquad \text{(XII.1)}$$

has a solution in $\bar{\Omega} \cap \text{dom } L$ *if and only if there exists a linear one-to-one mapping* Λ : coker $L \to$ ker L *such that the operator*

$$\tilde{M}_\Lambda = R_\Lambda P + (\Lambda\Pi + K_{P,Q})N \qquad \text{(XII.2)}$$

has a fixed point in $\bar{\Omega}$, *where* R_Λ : ker $L \to$ ker L *is any projector such that*

$$\text{Im } R_\Lambda = \text{Im } \Lambda \; .$$

Proof. Necessity. If $x \in \text{dom } L \cap \bar{\Omega}$ is a solution of (XII.1) then it follows from Proposition III.0 that

$$x = \tilde{M}_\Lambda x$$

with

$$\tilde{M}_\Lambda = P + (\tilde{\Lambda}\Pi + K_{P,Q})N$$

for any linear Λ : coker $L \to$ ker L which is one-to-one. Now let Y be any subspace of ker L of dimension equal to dim coker L and containing Px (such a subspace necessarily exists) and let R_Y be any projector in ker L such that Im $R_Y = Y$.

hen necessarily

$$Px = R_Y\, Px$$

nd if we take Λ_Y : coker L → ker L one-to-one such that Im Λ_Y = Y (such a linear apping necessarily exists) then

$$x = M_{\Lambda_Y}\, x = Px + (\Lambda_Y \Pi + K_{P,Q})Nx$$
$$= R_Y\, Px + (\Lambda_Y \Pi + K_{P,Q})Nx = \tilde{M}_{\Lambda_Y}\, x.$$

<u>Sufficiency</u>. If

$$x = R_\Lambda\, Px + (\Lambda\Pi + K_{P,Q})Nx$$

hen

$$(I - P)x = K_{P,Q}\, Nx$$

$$Px = R_\Lambda\, Px + \Lambda\Pi Nx$$

nd hence

$$Lx = (I - Q)Nx$$

$$(I - R_\Lambda)Px = \Lambda\Pi Nx = R_\Lambda\, \Lambda\Pi Nx$$

hich implies

$$Lx = (I - Q)Nx$$

$$(I - R_\Lambda)Px = 0 \ , \ R_\Lambda\, \Lambda\Pi Nx = 0$$

nd hence

$$Lx = Nx \ , \ Px = R_\Lambda\, Px \ .$$

hus x is a solution to (XII.2) such that $Px \in \mathrm{Im}\,\Lambda$.

Now assume that

$$0 \notin (L - N)(\mathrm{dom}\, L \cap \partial\Omega)$$

nd let $Y \subset \ker L$ be a vector subspace such that

$$\dim Y = \dim \operatorname{coker} L \ .$$

hen the Leray-Schauder degree

$$d\,[\,I - \tilde{M}_Y, \Omega, 0\,]$$

ith

$$\tilde{M}_Y = R_Y\, P + (\Lambda_Y\, \Pi + K_{P,Q})N \ ,$$

R_Y : ker L → ker L a projector such that

$$\mathrm{Im}\, R_Y = Y$$

and Λ_Y : coker L → ker L an orientation preserving one-to-one linear mapping such that

$$\mathrm{Im}\, \Lambda_Y = Y$$

is well-defined and it can be proved exactly like in chapter III that this number depends only upon L,N,Ω and Y. We shall denote it by

$$d_Y[(L,N),\Omega]$$

and call it the *coincidence degree of* L *and* N *in* Ω *with respect to* Y and one at once gets a result corresponding to Theorems III.1 to III.3 for $d_Y[(L,N),\Omega]$. On course a result like Proposition XII.1 cannot be proved for $d_Y[(L,N),\Omega]$ because

$$I - \tilde{M}_Y = I - R_Y P - (\Lambda_Y \Pi + K_{P,Q})N =$$

$$= (I - P - K_{P,Q} N) + (I - R_Y)P - \Lambda_Y \Pi N$$

has not a range necessarily contained in a proper subspace of X.

One also gets easily the following continuation theorem which corresponds essentially to Corollary IV.1 (in a less general form sufficient for applications).

Theorem XII.4. *Let* L *and* N *be like above (with* Ind L ≥ 0) *and assume that the following conditions are satisfied for some subspace* $Y \subset$ ker L *such that* dim Y = dim coker L.

(1) $Lx \neq \lambda Nx$ *for each* $x \in$ dom L $\cap\, \partial\Omega$ *and* $\lambda \in\,]0,1[$

(2) $QNx \neq 0$ *for each* $x \in$ ker L $\cap\, \partial\Omega$ *and* Q : Z → Z *a projector in* Z *such that* ker Q = Im L.

(3) $d[JQN|Y, \Omega \cap Y, 0] \neq 0$ *where* J : Im Q → Y *is some isomorphism.*

Then equation (XII.1) *has at least one solution* $x \in$ dom L $\cap\, \overline{\Omega}$.

It is clear now that most of the results of Chapter VII can be adapted to that situation. In particular one has the following result which corresponds to Theorem VII.4.

Theorem XII.5. *Assume that the following conditions hold.*

(a) *There exists* $\delta \in [0,1[$, $\mu \geq 0$, $\nu \geq 0$ *such that, for each* $x \in X$,

$$|K_{P,Q} Nx| \leq \mu |x|^{\delta} + \nu .$$

(b) $(\forall$ *bounded* $V \subset \ker P)(\exists\ t_o > 0)(\forall t > t_o)(\forall\ z \in V)$

$(\forall\ w \in Y \cap \partial B(1)) : QN(tw + t^{\delta} z) \neq 0.$

(c) *For some* $t \geq t_o$, $d[JQN|Y, B(t), 0] \neq 0$ *with* $J : \operatorname{Im} Q \to Y$ *an isomorphism.*

Then $(L - N)(\operatorname{dom} L) \supset \operatorname{Im} L$.

As an application let us consider now the elliptic problem treated in Chapter III section 7 but under the more general assumption that the elliptic operator L ith the given boundary conditions is a Fredholm operator with nonnegative index. et us make the assumptions of Section 7 in Chapter VIII on the nonlinearity f nd replace the assumption (UC) by

(H) *There exists a linear one-to-one mapping*

$$T : \operatorname{Im} Q \to \ker L$$

uch that if

$$P_T(c) = \sup_{z \in \operatorname{Im} Q} \operatorname{meas} \{x \in \Omega ; \frac{|Tz(x)|}{|z(x)|} < c \}$$

vith $0/0 = \infty$) *then*

$$P_T(c) \to 0 \quad \text{if} \quad c \to 0.$$

Now let $w'_1, \ldots, w'_{d*}$ be smooth functions spanning Im Q and if $a = (a_1, \ldots, a_{d*})$, et us write $a.w'$ for $a_1 w'_1 + \ldots + a_{d*} w'_{d*}$. Define $X : R^{d*} \to R^{d*}$ by

$$_j(a) = \int_{\{x \in \Omega : T(a.w'(x)) > 0} h_+(x) w'_j(x)\, dx + \int_{\{x \in \Omega : T(a.w'(x)) < 0} h_-(x) w'_j(x)\, dx$$

$$(j = 1, 2, \ldots, d^*)$$

t follows then from assumption (H) and a reasoning similar to that of Lemma VIII.1 hat

$$X_j(a) = \lim_{r \to \infty} \int_{\Omega} f(x, r(a.Tw'(x)) + u(x))\, w'_j(x)\, dx$$

$$(j = 1, 2, \ldots, d^*)$$

niformly for u bounded in $L^1(\Omega)$ and $a \in S^{d*-1}$. In particular X_j is continuous on S^{d*-1} and proceeding like in section 7 of chapter VIII but through Theorem XII.5 with Y the d*-dimensional subspace of ker L spanned by

$$Tw'_1, \dots, Tw'_{d*} ,$$

on gets the following

Theorem XII.6. *If the assumptions above hold and if*

$$X(a) \neq 0 \text{ for } a \in S^{d*-1}$$

and

$$d[\tilde{X}, B(1), 0] \neq 0$$

with $\tilde{X}$ *any continuous extension to* $B(1) \subset R^{d*}$ *of* X, *then problem* (VIII.16) *has at least one solution.*

Corollary XII.2. *If the assumptions above hold and if*

$$M_T(z) > 0$$

for all $z \in \text{Im } Q$ *with*

$$M_T(z) = \int_{Tz > 0} h_+ z + \int_{Tz > 0} h_- z ,$$

then problem (VIII.16) *has at least one solution.*

3. Another way to circumvent the Proposition XII.1 is to use stronger topological tools than degree theory. Let X be a normed space, $R : X \to X$ a continuous projector with range of dimension $m > 0$,

$$\Phi : \text{Im } R \cap \overline{B(r)} \to Y \subset \text{Im } R$$

a continuous mapping such that

$$0 \notin \Phi(\text{Im } R \cap \partial B(r))$$

with Y a vector subspace of Im R of dimension $p < m$. Let S^{k-1} denote the unit sphere in R^k and let

$$H : S^{m-1} \to \text{Im } R \cap \partial B(1)$$

$$H^1 : Y \cap \partial B(1) \to S^{p-1}$$

be isomorphisms. Lastly let $\psi : S^{m-1} \to S^{p-1}$ be defined by

$$\Psi(u) = H' \frac{\Phi(rHu)}{|\Phi(rHu)|}$$

We give without proof the following.

Proposition XII.4. *Suppose that* ψ *has non trivial stable homotopy. Then every mapping* $F : \overline{B(r)} \to X$ *such that* $F = \tilde{F}(.,1)$ *for some compact mapping*

$$\tilde{F} : \overline{B(r)} \times [0,1] \to X$$

verifying

$$x \neq F(x,\lambda) \text{ for } x \in \partial B(r),\ \lambda \in [0,1]$$

and

$$F(.,0) = R + \Phi R$$

has at least one fixed point in $B(r)$.

Recall that having non trivial stable homotopy for ψ means the following. If $\psi : S^{m-1} \to S^{p-1}$ is continuous one can define a mapping

$$\Sigma\psi : S^m \to S^p$$

which is called the *suspension* of ψ and which is constructed as follows. If one thinks S^{m-1} (resp. S^{p-1}) as the equator of S^m(resp. S^p) then $\Sigma\psi$ will map the north (resp. south) pole of S^m onto the north (resp. south) pole of S^p; of γ is a half greatcircle on S^m joining the poles and x its intersection with the equator, and if γ' is the half greatcircle on S^p joining the poles and passing through $\psi(x)$, then $\Sigma\psi$ is defined on γ as a map from γ to γ' linear with respect to the arc length. Thus $\Sigma\psi$ is clearly a continuous extension of ψ which maps S^m into S^p. Now it is clear that if ψ and ψ' are homotopically equivalent mappings of S^{m-1} into S^{p-1}, then $\Sigma\psi$ and $\Sigma\psi'$ are homotopically equivalent. However $\Sigma\psi$ and $\Sigma\psi'$ can be homotopically equivalent even if ψ and ψ' are not. However if the processus of suspension for ψ is iterated one can prove that for $j \geq m - p$, $\Sigma^{j+1}\psi$ is homotopically nontrivial (i.e. is not homotopic to a constant map from S^{m+j} into S^{p+j}) if and only if $\Sigma^j\psi$ is homotopically nontrivial. In this case ψ is said to have *nontrivial stable homotopy*.

Using Proposition XII.4 instead of Leray-Schauder degree one can then prove in a similar way as for Corollary IV.1 the following.

Theorem XII.7. *Let* X,Z *be normed spaces,* $L : \text{dom } L \subset X \to Z$ *a linear Fredholm mapping with* $\text{Ind } L > 0$, $N : \text{cl } B(r) \subset X \to Z$ *be L-compact and suppose that the following conditions are satisfied* :

(1) $Lx \neq \lambda Nx$ *for each* $x \in \partial B(r)$ *and* $\lambda \in]0,1[$;

(2) $QNx \neq 0$ *for each* $x \in \partial B(r) \cap \ker L$

(3) *the following mapping* Ψ *has nontrivial stable homotopy*,

$$\Psi : S^{m-1} \to S^{p-1} \quad , u \to H'\left(\frac{QN(rHu)}{|QN(rHu)|}\right) \tag{XII.3}$$

with $m = \dim \ker L$, $p = \dim \operatorname{coker} L$, $H : S^{m-1} \to \ker L \cap \partial B(1)$, $H' : \operatorname{Im} Q \cap \partial B(1) \to S^{p-1}$ *are isomorphisms.*

Then equation (XII.1) *has at least one solution in* $\overline{B(r)}$.

From this result one gets easily the following one, which corresponds to Theorem VII.4.

Theorem VII.8. *Assume that instead of* (1)-(2)-(3) *in Theorem* VII.7 *the conditions* (a)-(b) *of Theorem* XII.5 *with* $Y = Z$ *and*

(c') *the mapping* Ψ *defined in* (XII.3) *with* $r \geq t_o$ *has non trivial stable homotopy.*

Then

$$(L - N)(\operatorname{dom} L) \supset \operatorname{Im} L .$$

This theorem can then be used as in Chapter VIII to extend Theorem VIII.3 to the case where $\operatorname{Ind} L > 0$ when the assumption of the non-vanishing of the degree is replaced by assuming that the mapping $\Psi : S^{d-1} \to S^{d*-1}$ (with $d = \dim \ker$ $d* = \dim \operatorname{coker} L$) defined by $\Psi = \phi/|\phi|$ with

$$\phi_i(a) = \int_{a.w > 0} h_+(x)w'_i(x)dx + \int_{a.w < 0} h_-(x)w'_i(x)dx \;, \qquad (i = 1,\ldots,d*)$$

$a.w = \sum_{i=1}^{d} a_i w_i$ and $\{w_1,\ldots,w_d\}$ (resp. $\{w'_1,\ldots,w'_{d*}\}$) span $\ker L \cap \ker B$ (resp. $(\operatorname{Im} L \cap \ker B)$), has nontrivial stable homotopy.

Bibliographical notes concerning Chapter XII

Proposition XII.1 can be found in Nirenberg (*Troisième Colloque du CBRM d'analyse fonctionnelle*, Vander, Louvain, 1971, 57-74) and in Mawhin (*J. Different Equations* 12 (1972) 610-636). The theorems of section 2 are an abstract formulatio of results due to Rabinowitz (*Indiana University Mathematics Journal* 22 (1972) 43-49) where one can found the application to the elliptic problem sketched at the end of the section. The concept of genus is due to Krasnosel'skii ("Topological

Methods in the theory of nonlinear integral equations", 1963) and the present approach due to Coffman (*J. d'Anal. Math.* 22 (1969) 391-419) can also be found in Rabinowitz (*Rocky Mountain J. of Math.* 3 (1973) 161-202 ; "Théorie du degré topologique et applications à des problèmes aux limites non linéaires", Paris, 1975). Proposition XII.3 and Theorems XII.4 and 5 are given here for the first time and were inspired by Theorem XII.6 and Corollary XII.2 which were given by Schechter (*An.Scuola Norm. Sup. Pisa* (3) 27 (1973) 707-716) with another proof. Proposition XII.4 is due to Nirenberg (op. citae) whose work has inspired Theorems XII.7 and XII.8, the application to the elliptic problem being the problem originally treated by Nirenberg. For further work in this line see Tromba (Stable homotopy groups of spheres and non-linear P.D.E, *to appear*), Cronin (*J. Differential Equations* 14 (1973) 581-596) Berger and Podolak (*Bull Amer. Math. Soc.* 80 (1974) 861-864). Ize (Courant Institute New-York Univ. Ph. D thesis, 1974. For related work about the generalized topological invariants see Švarc (*Soviet Math. Dokl.* 5 (1964) 57-59), Geba (*Fund. Math.* 54 (1964) 177-209; 64 (1969) 341-373), Geba and Granas (*J. Math. Pures Appl.* 52 (1973) 145-270), Smale (*Amer. J. Math.* 87 (1965) 861-866), Elworthy and Tromba (*Proc. Symp. Pure Math.* 15 (1970) 86-99), Quinn (*ibid.* 15, 213-227), Granas (*Sémin. Collège de France*, 1969-70), Borisorich and Sapronov (*Soviet Math. Dokl.* 12 (1971) 5-9), Fenn (Generalized Leray-Schauder degrees, *to appear*), Mukherjea (*J. Math Mechanics* 19 (1970) 731-744, Isnard (Ph. D. Thesis, Univ. of Chicago, 1972).

For a general references concerning this chapter, see Nirenberg ("Topics in Nonlinear Functional Analysis" 1974).

REFERENCES.

V.G. ADAMOV and N.A. BOBYLEV, A boundary value problem, Differential Equations 10 (1974) 891-894. (Ch. V).

H. AMANN, Existence of multiple solutions for nonlinear elliptic boundary value problems, Indiana Univ. Math. J. 21 (1972) 925-935.

H. AMANN, Fixed points of asymptotically linear maps in ordered Banach spaces, J. Functional Analysis 14 (1973) 162-171.

H. AMANN, "Lectures on some Fixed Point Theorems", Monografias de Matematica, IMPA, Rio de Janeiro, s.d. (Ch. III, IV).

H.A. ANTOSIEWICZ, Boundary value problems for nonlinear ordinary differential equations, Pacific J. Math. 17 (1966) 191-197.

P. BAILEY, L. SHAMPINE and P. WALTMAN, "Nonlinear Two Point Boundary Value Problems", Academic Press, New York, 1968.

S. BANCROFT, Perturbations with several independent parameters, J. Math. Anal. Appl. 50 (1975) 383-414. (Ch. II).

S. BANCROFT, J.K. HALE and D. SWEET, Alternative problems for nonlinear functional equations, J. Differential Equations 4 (1968) 40-56.

C. BANFI, Su un metodo di successive approssimazioni per lo studio delle soluzioni periodiche di sistemi debolmente nonlineari, Atti Accad. Sci. Torino Cl. Sci. Fis. Mat. Natur. 100 (1965-66) 471-479.

R.G. BARTLE, Singular points of functional equations, Trans. Amer. Math. Soc. 75 (1953) 366-384.

J.W. BEBERNES, A simple alternative problem for finding periodic solutions of second order ordinary differential systems, Proc. Amer. Math. Soc. 42 (1974) 121-127.

J.W. BEBERNES, R.E. GAINES and K. SCHMITT, Existence of periodic solutions for third and fourth order ordinary differential equations via coincidence degree, Ann. Soc. Sci. Bruxelles Ser. I 88 (1974) 25-36.

J. BEBERNES and K. SCHMITT, An existence theorem for periodic boundary value problems for systems of second order differential equations, Arch. Math. (Brno) (4) 8 (1972) 173-176.

J.W. BEBERNES and K. SCHMITT, Periodic boundary value problems for systems of second order differential equations, J. Differential Equations 13 (1973) 32-47.

M. BERGER and M. BERGER, "Perspectives in Nonlinearity", Benjamin, New York, 1968.

M.S. BERGER and E. PODOLAK, On nonlinear Fredholm operator equations, Bull. Amer. Math. Soc. 80 (1974) 861-864.

S.R. BERNFELD and V. LAKSHMIKANTHAM, "An Introduction to Nonlinear Boundary Value Problems", Academic Press, New York, 1974.

S. BERNFELD, G.S. LADDE and V. LAKSHMIKANTHAM, Existence of solutions of two point boundary value problems for nonlinear systems, J. Differential Equations 18 (1975) 103-110. (Ch. V).

Yu.G. BORISOVICH and Yu.I. SAPRONOV, A contribution to the topological theory of condensing operators, Soviet Math. Dokl. 9 (1968) 1304-1308.

Yu.G. BORISOVICH and Yu.I. SAPRONOV, On some topological invariants of non-linear Fredholm mappings, Soviet Math. Dokl. 12 (1971) 5-9.

H. BREZIS, Une équation elliptique non-linéaire à la résonance, Séminaire Lions-Schwartz, 1971.

H. BREZIS, Quelques propriétés des opérateurs monotones et des semi-groupes non-linéaires, NATO Summer School on Nonlinear Operators and the Calculus of Variations, Bruxelles, 1975, to appear. (Ch. VII, VIII).

H. BREZIS and A. HARAUX, Sur l'image d'une somme d'opérateurs et applications, Israel J. Math. 1976, to appear. (Ch. VII, VIII).

H. BREZIS and L. NIRENBERG, On some nonlinear operators and their ranges, Ann. Sc. Norm. Sup. Pisa, to appear. (Ch. VII, VIII).

F.E. BROWDER, Topology and non-linear functional equations, Studia Math. 31 (1968) 189-204.

F.E. BROWDER, "Nonlinear operators and nonlinear equations of evolution in Banach spaces", Proceed. Symposia Pure Math. 18, 2, Amer. Math. Soc., Providence, 1976.

F.E. BROWDER and R.D. NUSSBAUM, The topological degree for noncompact nonlinear mappings in Banach spaces, Bull. Amer. Math. Soc. 74 (1968) 671-676.

R. CACCIOPPOLI, Sulle corrispondenze funzionali inverse diramate : teoria generale e applicazioni ad alcune equazioni funzionali non lineari e al problema di Plateau, Atti Accad. Naz. Lincei Rend. Cl. Sci. Fis. Mat. Natur. Sez. I 24 (1936) 258-263 ; 416-421.

L. CESARI, Sulla stabilità delle soluzioni dei sistemi di equazioni differenziali lineari a coefficienti periodici, Atti Accad. Ital. Mem. Cl. Fis. Mat. Nat. (6) 11 (1940) 633-692.

L. CESARI, Periodic solutions of hyperbolic partial differential equations, in "Intern. Symposium on Nonlinear Differential Equations and Non-linear Mechanics", Academic Press, New York, 1963, 33-57.

L. CESARI, Functional analysis and periodic solutions of nonlinear differential equations, Contributions to Differential Equations 1 (1963) 149-187.

L. CESARI, Functional analysis and Galerkin's method, Michigan Math. J. 11 (1964) 385-414.

L. CESARI, Nonlinear Analysis, in "Nonlinear Mechanics", C.I.M.E. Lecture Notes, Cremonese, Roma, 1973, 3-95.

L. CESARI, Alternative methods in nonlinear analysis, in "Intern. Conf. Differential Equations", H.A. Antosiewicz ed., Academic Press, 1975, 95-148. (Ch. II).

L. CESARI, Nonlinear oscillations in the frame of alternative methods, in "Dynamical Systems : An International Symposium", vol. I, Academic Press, New York, 1976, 29-50. (Ch. II-IX).

L. CESARI and R. KANNAN, Functional analysis and nonlinear differential equations, Bull. Amer. Math. Soc. 79 (1973) 1216-1219. (Ch. II).

L. CESARI and R. KANNAN, Periodic solutions in the large of nonlinear ordinary differential equations, Rend. di Matem. (6) 8 (1975) 633-654. (Ch. V, IX).

S.H. CHANG, Periodic solutions of certain second order nonlinear differential equations, J. Math. Anal. Appl. 49 (1975) 263-266. (Ch. IX).

S.H. CHANG, Periodic solutions of $(a(t)x')' + f(t,x) = p(t)$, Quart. J. Math. Oxford (2) 27 (1976) 105-109. (Ch. IX).

H.C.C. CHEN, "Constructive Methods for Nonlinear Boundary Value Problems", Ph.D. Thesis, Colorado State University, Fort Collins, 1974.

C.V. COFFMAN, A minimum-maximum principle for a class of nonlinear integral equations, J. Analyse Math. 22 (1969) 391-419.

COLE, "Theory of Ordinary Differential Equations", Century, Crofts, New York, 1968.

C. CORDUNEANU, Sisteme diferentiale care admit solutii marginite, Acad. R.P. Rouman. Fil. Iasi, Studi Cerc. Stunt. Mat. 8 (1957) 107-126.

C. CORDUNEANU, Sopra i problemi ai limiti per alcuni sistemi di equazioni differenziali non lineari, Rend. Accad. Sci. Fis. Mat. Soc. Naz. Sci. Lett. Arti Napoli (4) 25 (1958) 3-11.

M.G. CRANDALL and P.H. RABINOWITZ, Bifurcation from simple eigenvalues, J. Functional Analysis 8 (1971) 321-340. (Ch. X, XI).

J. CRONIN, Branch points of solutions of equations in Banach spaces, Trans. Amer. Math. Soc. 69 (1950) 208-231 ; 76 (1954) 207-222.

J. CRONIN, A definition of degree for certain mappings in Hilbert space, Amer. J. Math. 73 (1951) 763-772.

. CRONIN, "Fixed Points and Topological Degree in Nonlinear Analysis", Amer. Math. Soc., Providence, 1964.

. CRONIN, Functional equations with several solutions, J. Math. Anal. Appl. 28 (1969) 159-173. (Ch. X).

. CRONIN, Periodic solutions of nonautonomous equations, Boll. Un. Mat. Ital. (4) 6 (1972) 45-54.

. CRONIN, Equations with bounded nonlinearities, J. Differential Equations 14 (1973) 581-596.

. CRONIN, Quasilinear equations and equations with large nonlinearities, Rocky Mountain J. Math. 4 (1974) 41-63. (Ch. V, IX).

.M. CUSHING, Some existence theorems for nonlinear eigenvalue problems associated with elliptic equations, Arch. Rat. Mech. Anal. 42 (1971) 63-76.

.N. DANCER, On a nonlinear elliptic boundary-value problem, Bull. Austral. Math. Soc. 12 (1975) 399-405. (Ch. VIII).

.N. DANCER, On the Dirichlet problem for weakly nonlinear elliptic partial differential equations, to appear. (Ch. VIII).

. DANES, On densifying and related mappings and their application in nonlinear functional analysis, in "Theory of Nonlinear Operators, Proceedings of a Summer-School", Akademie-Verlag, Berlin, 1974, 15-56.

. DARBO, Punti uniti in transformazioni a codominio non compatto, Rend. Sem. Mat. Univ. Padova 24 (1955) 84-92.

.G. DE FIGUEIREDO, Some remarks on the Dirichlet problem for semilinear elliptic equations, Univ. de Brasilia, Trabalho de Matematica n° 57, 1974.

.G. DE FIGUEIREDO, On the range of nonlinear operators with linear asymptotes which are not invertible, Comment. Math. Univ. Carolinae 15 (1974) 415-428.

.G. DE FIGUEIREDO, The Dirichlet problem for nonlinear elliptic equations : A Hilbert space approach, in "Partial Differential Equations and related topics", Lecture Notes n° 446, Springer, Berlin, 1975, 144-165.

. DEIMLING, "Nichtlineare Gleichungen und Abbildungsgrad", Springer, Berlin, 1974. (Ch. III, IV).

. DE SIMON and G. TORELLI, Soluzioni periodiche di equazioni a derivate parziali di tipo iperbolico non lineari, Rend. Sem. Univ. Padova 40 (1968) 380-401.

. DUBROVSKII, Sur certaines équations intégrales non linéaires, Ucen. Zap. Moskow. Gos. Univ. 30 (1939) 49-60.

K.D. ELWORTHY and A.J. TROMBA, Differential structures and Fredholm maps on Banach manifolds, Proc. Symp. Pure Math. 15 (1969) 45-94.

K.D. ELWORTHY and A.J. TROMBA, Degree theory on Banach manifolds, Proc. Symp. Pure Math. 18 (1) (1970) 86-94.

H. EPHESER, Uber die Existenz der Lösungen von Randwertaufgaben mit gewöhnlichen, nichtlinearen Differentialgleichungen zweiter Ordnung, Math. Zeits. 61 (1955) 435-454.

L.H. ERBE, Nonlinear boundary value problems for second order differential equations, J. Differential Equations 7 (1970) 459-472.

J.O.C. EZEILO, On the existence of periodic solutions of a certain third order differential equation, Proc. Cambridge Philos. Soc. 56 (1960) 381-389.

C. FABRY, Weakly nonlinear equations in Banach spaces, Boll. Un. Mat. Ital. (4) 4 (1971) 687-700.

C. FABRY and C. FRANCHETTI, Teoremi di esistenza per equazioni non lineari in spazi di Banach e applicazioni, Univ. di Firenze, Seminari dell'Istituto di Matematica Applicata, 1974.

C. FABRY and C. FRANCHETTI, Nonlinear equations with growth restrictions on the nonlinear terms, J. Differential Equations 20 (1976) 283-291.

R. FENN, Generalized Leray-Schauder degrees, to appear.

R.E. FENNELL, Periodic solutions of functional differential equations, J. Math. Anal. Appl. 39 (1972) 198-201.

R. FIORENZA, Sulla hölderianità delle soluzioni dei problemi di derivata obliqua regolare del secondo ordine, Ricerche Mat. 14 (1965) 102-123.

C. FRANCHETTI, Perturbations non bornées de l'oscillateur dans le cas de résonance, Univ. di Firenze, Seminari dell'Istituto di Matematica Applicata, 1973.

K.O. FRIEDRICHS, "Special Topics in Analysis", New York University, Lecture Notes, New York, 1953-54.

S. FUCIK, Further remark on a theorem by E.M. Landesman and A.C. Lazer, Comment. Math. Univ. Carolinae 15 (1974) 259-271.

S. FUCIK, Nonlinear equations with noninvertible linear part, Czechoslovak Math. J. 24 (99) (1974) 467-495.

S. FUCIK, Surjectivity of operators involving linear noninvertible part and nonlinear compact perturbations, Funkcial. Ekvac. 17 (1974) 73-83.

S. FUCIK, Boundary value problems with jumping nonlinearities, Casopis Pest. Matem. 101, to appear.

. FUCIK, M. KUCERA and J. NECAS, Ranges of nonlinear asymptotically linear operators, J. Differential Equations 17 (1975) 375-394.

. FUCIK and J. MAWHIN, Periodic solutions of some nonlinear differential equations of higher order, Casopis Pest. Mat. 100 (1975) 276-283.

. FUCIK, J. NECAS, J. SOUCEK, Vl. SOUCEK, "Spectral Analysis of Nonlinear Operators", Lecture Notes in Math. n° 346, Springer, Berlin, 1973. (Ch. III, IV).

.E. GAINES, A priori bounds and upper and lower solutions for nonlinear second-order boundary value problems, J. Differential Equations 12 (1972) 291-312.

.E. GAINES, A priori bounds for solutions to nonlinear two-point boundary value problems, Applicable Analysis 3 (1973) 157-167. (Ch. V).

.E. GAINES, Difference equations associated with boundary value problems for second order nonlinear ordinary differential equations, SIAM J. Numer. Anal. 11 (1974) 411-434. (Ch. V).

.E. GAINES, Existence of periodic solutions to second-order nonlinear ordinary differential equations, J. Differential Equations 16 (1974) 186-199.

.E. GAINES and J. MAWHIN, Ordinary differential equations with nonlinear boundary conditions, to appear.

. GEBA, Algebraic topology methods in the theory of compact fields in Banach spaces, Fund. Math. 54 (1964) 177-209.

. GEBA, Fredholm σ-proper maps of Banach spaces, Fund. Math. 64 (1969) 341-373.

. GEBA and A. GRANAS, Infinite dimensional cohomology theories, J. Math. Pures Appl. 52 (1973) 145-270.

.H. GEORGE and W.G. SUTTON, Application of Liapunov theory to boundary value problems, Proc. Amer. Math. Soc. 25 (1970) 666-671.

.H. GEORGE and R.J. YORK, Application of Liapunov theory to boundary value problems II, Proc. Amer. Math. Soc. 37 (1973) 207-212. (Ch. V).

.S. GOLDENSTEIN, I. Ts. GOKHBERG and A.S. MARKUS, Investigation of some properties of bounded linear operators in connection with their q-norm, Uch. Zap. Kishinev. Gos. Univ. 29 (1957) 29-36.

. GRANAS, On a certain class of non-linear mappings in Banach spaces, Bull. Acad. Polon. Sci. Sér. Sci. Math. Astronom. Phys. 9 (1957) 867-871.

. GRANAS, The theory of compact vector fields ans some of its applications to topology of functional spaces (I), Rozpravy Mat. 20 (1962) 1-93.

. GRANAS, "Topics in infinite dimensional topology", Séminaire sur les équations aux dérivées partielles, Collège de France, Paris, 1969-1970.

V.V. GUDKOV and A.Ya. LEPIN, The solvability of certain boundary value problems for ordinary second order differential equations, Differential Equations 7 (1971) 1779-1788. (Ch. V).

V.V. GUDKOV and A.Ya. LEPIN, On necessary and sufficient conditions for the solvability of certain boundary value problems for a second-order ordinary differential equation, Soviet Math. Dokl. 14 (1973) 800-803. (Ch. V).

V.V. GUDKOV, Io.A. KLOKOV, A.Ya. LEPIN and V.D. PONOMAREV, "Two-point Boundary Value Problems for Ordinary Differential Equations" (Russian), Izdat. "Zinatne", Riga, URSS, 1973. (Ch. V).

C.P. GUPTA, Sum of ranges of operators and nonlinear elliptic boundary value problems, to appear. (Ch. VII, VIII).

G. GUSSEFELDT, Der topologische Abbildungsgrad für vollstetige Vektorfelder zum Nachweis von periodische Lösungen, Math. Nachr. 36 (1968) 231-233.

G.B. GUSTAFSON and K. SCHMITT, Periodic solutions of hereditary differential systems, J. Differential Equations 13 (1973) 567-587.

G.B. GUSTAFSON and K. SCHMITT, A note on periodic solutions for delay-differential systems, Proc. Amer. Math. Soc. 42 (1974) 161-166.

K. GUSTAFSON and D. SATHER, Large nonlinearities and monotonicity, Arch. Rat. Mech. Anal. 48 (1972) 109-122.

K. GUSTAFSON and D. SATHER, Large nonlinearities and closed linear operators, Arch. Rat. Mech. Anal. 52 (1973) 10-19. (Ch. II).

A. HALANAY, Solutions périodiques des systèmes non-linéaires à petit paramètre, Atti Accad. Naz. Lincei Rend. Cl. Sci. Fis. Mat. Natur. (8) 22 (1957) 30-32.

J.K. HALE, Periodic solutions of non-linear systems of differential equations, Riv. Mat. Univ. Parma 5 (1954) 281-311.

J.K. HALE, "Ordinary Differential Equations", Wiley, New York, 1969.

J.K. HALE, "Functional Differential Equations", Springer, New York, 1971.

J.K. HALE, "Applications of Alternative Problems", Brown University Lecture Notes 71-1, Providence, 1971.

J.K. HALE and J. MAWHIN, Coincidence degree and periodic solutions of neutral equations, J. Differential Equations 15 (1974) 295-307.

W.S. HALL, Periodic solutions of a class of weakly nonlinear evolution equations, Arch. Rat. Mech. Anal. 39 (1970) 294-332.

W.S. HALL, On the existence of periodic solutions for the equation $D_{tt}u + (-1)^p D_x^{2p} u = \varepsilon f(t,x,u)$, J. Differential Equations 7 (1970) 509-526.

.S. HALL, The bifurcation of solutions in Banach spaces, Trans. Amer. Math. Soc. 161 (1971) 207-218.

.A. HARRIS, Holomorphic solutions of nonlinear differential equations at singular points, in "Advances in Differential and Integral Equations", SIAM Symp. Madison, Wisc., 1969, 184-187.

.A. HARRIS, Jr., Y. SIBUYA and L. WEINBERG, Holomorphic solutions of linear differential systems at singular points, Arch. Rat. Mech. Anal. 35 (1969) 245-248.

. HARTMAN, On boundary value problems for systems of ordinary, nonlinear, second order differential equations, Trans. Amer. Math. Soc. 96 (1960) 493-509.

. HARTMAN, On two-point boundary value problems for nonlinear second order systems, SIAM J. Math. Anal. 5 (1974) 172-177. (Ch. V).

.W. HEIDEL, A second order nonlinear boundary value problem, J. Math. Anal. Appl., to appear. (Ch. V).

. HESS, On a theorem by Landesman and Lazer, Indiana Univ. Math. J. 23 (1974) 827-830.

. HETZER, Some remarks on $\emptyset_+$ operators and on the coincidence degree for a Fredholm equation with noncompact nonlinear perturbation, Ann. Soc. Sci. Bruxelles Ser. I 89 (1975) 497-508.

. HETZER, Some applications of the coincidence degree for set-contractions to functional differential equations of neutral type, Comm. Math. Univ. Carolinae 16 (1975) 121-138.

. HETZER, On the existence of weak solutions for some quasilinear elliptic variational boundary value problems at resonance, Comm. Math. Univ. Carolinae 17 (1976) 315-334.

. HETZER and V. STALLBOHM, Eine Existenzaussage für asymptotisch lineare Störungen eines Fredholmoperators mit index zero, to appear.

. HETZER and V. STALLBOHM, Coincidence degree and Rabinowitz's bifurcation theorem, Publ. Inst. Math. Beograd 20 (34) (1976).

. HETZER and V. STALLBOHM, Global behavior of bifurcation branches and the essential spectrum, to appear.

.A.S. ISNARD, "Degree Theory in Banach Manifolds", Univ. Chicago Ph.D. Thesis, 1972.

.A. IZE, "Bifurcation Theory for Fredholm Operators", New York University Ph.D. Thesis, 1974.

.K. JACKSON, Subfunctions and second-order ordinary differential inequalities, Advances in Math. 2 (1968) 307-363.

L.K. JACKSON, A Nagumo condition for ordinary differential equations, Proc. Amer. Math. Soc., to appear. (Ch. V).

M. JEPPSON, SIAM Mathematical Topics in Economic Theory and Computation, 1972, 122-129.

R.I. KACHUROWSKII, On Fredholm theory for nonlinear operator equations, Soviet Math. Dokl. 11 (1970) 751-754.

R.I. KACHUROWSKII, On nonlinear operators whose ranges are subspaces, Soviet Math. Dokl. 12 (1971) 168-172.

R. KANNAN, Existence of periodic solutions of nonlinear differential equations, Trans. Amer. Math. Soc. 217 (1976) 225-236.

R. KANNAN and J. SCHUUR, Boundary value problems for even order nonlinear ordinary differential equations, Bull. Amer. Math. Soc. 82 (1976) 80-82. (Ch. V).

J.L. KAPLAN, A. LASOTA and J.A. YORKE, An application of the Wazewski retract method to boundary value problems, Zeszyty Nauk Uniw. Jagielle, to appear. (Ch. V).

J.L. KAZDAN and F.W. WARNER, Remarks on some quasilinear elliptic equations, Comm. Pure Appl. Math., to appear. (Ch. VIII).

I.T. KIGURADZE, A priori estimates for derivatives of bounded functions satisfying second-order differential inequalities, Differential Equations 3 (1967) 541-546. (Ch. V).

I.T. KIGURADZE, A singular two-point boundary value problem, Differential Equations 5 (1969) 1493-1504.

I.T. KIGURADZE, Some singular boundary value problems for ordinary nonlinear second order differential equation, Diff. Equat. 4 (1968) 901-910. (Ch. V).

K. KLINGELHÖFER, Nonlinear boundary value problems with simple eigenvalue of the linear part, Arch. Rat. Mech. Anal. 37 (1970) 382-398.

H.W. KNOBLOCH, Remarks on a paper by L. Cesari on functional analysis and nonlinear differential equations, Michigan Math. J. 10 (1963) 417-430.

H.W. KNOBLOCH, Eine neue Methode zur Approximations periodischer Lösungen nichtlinearer Differentialgleichungen zweiter Ordnung, Math. Zeits. 82 (1963) 177-197.

H.W. KNOBLOCH, On the existence of periodic solutions for second order vector differential equations, J. Differential Equations 9 (1971) 67-85.

H.W. KNOBLOCH and K. SCHMITT, Nonlinear boundary value problems for systems of differential equations, Proc. R. Soc. Edinburgh, to appear. (Ch. V).

H. KNOLLE, Lotka-Volterra equations with continously retarded argument and external influence, to appear.

M.A. KRASNOSEL'SKII, Some problems of nonlinear analysis, Amer. Math. Soc. Transl. (2) 10 (1958) 345-409.

M.A. KRASNOSEL'SKII, "Topological Methods in the Theory of Nonlinear Integral Equations", Pergamon, Oxford, 1963.

M.A. KRASNOSEL'SKII, La théorie des solutions périodiques d'équations différentielles non autonomes, Russian Math. Surveys 21 (1966) 53-74.

M.A. KRASNOSEL'SKII, "The Operator of Translation Along Trajectories of Ordinary Differential Equations", Amer. Math. Soc., Providence, 1968.

M.A. KRASNOSEL'SKII and E.A. LIFSCHITS, Duality principles for boundary value problems, Soviet Math. Dokl. 8 (1967) 1236-1239. (Ch. V).

M.A. KRASNOSEL'SKII and E.A. LIFSCHITS, Some criteria for the existence of periodic oscillations in nonlinear systems, Automat. Remote Control 9 (1973) 1374-1377.

M.A. KRASNOSEL'SKII, G.M. VAINIKKO, P.P. ZABREIKO, Ya.B. RUTITSKII, V.Ya. STETSENKO, "Approximate solution of operator equations", Wolters-Noordhoff, Groningen, 1972.

C. KURATOWSKI, Sur les espaces complets, Fund. Math. 15 (1930) 301-309.

B. LALOUX, Indice de coincidence et bifurcations, in "Equations différentielles et fonctionnelles non linéaires", Hermann, Paris, 1973, 109-121.

B. LALOUX, "Indice de coïncidence et bifurcations", Ph.D. Thesis, Univ. Cath. Louvain, 1974.

B. LALOUX, On the computation of the coincidence index in the critical case, Ann. Soc. Sci. Bruxelles Ser. I 88 (1974) 176-182.

B. LALOUX and J. MAWHIN, Coincidence index and multiplicity, Trans. Amer. Math. Soc. 217 (1976) 143-162.

B. LALOUX and J. MAWHIN, Multiplicity, Leray-Schauder formula and bifurcation, J. Differential Equations, to appear.

E.M. LANDESMAN and A.C. LAZER, Nonlinear perturbations of linear elliptic boundary value problems at resonance, J. Math. Mech. 19 (1970) 609-623.

A. LASOTA and J.A. YORKE, Existence of solutions of two-point boundary value problems for nonlinear systems, J. Differential Equations 11 (1972) 509-518.

A.C. LAZER, On the computation of periodic solutions of weakly nonlinear differential equations, SIAM J. Appl. Math. 15 (1967) 1158-1170.

A.C. LAZER, On Schauder's fixed point theorem and forced second-order nonlinear oscillations, in "Proceed. U.S.-Japan Semin. Functional and Differential Equations", Benjamin, 1967, 473-478. (Ch. IX).

A.C. LAZER, On Schauder's fixed point theorem and forced second-order nonlinear oscillations, J. Math. Anal. Appl. 21 (1968) 421-425.

A.C. LAZER and D.E. LEACH, Bounded perturbations of forced harmonic oscillations at resonance, Ann. Mat. Pura Appl. (4) 82 (1969) 49-68.

A. LEBOW and M. SCHECHTER, Semigroups of operators and measures of noncompactness, J. Funct. Anal. 7 (1971) 1-26.

M. LEES and M.H. SCHULTZ, A Leray-Schauder principle for A-compact mappings and the numerical solution of non-linear two-point boundary value problems, in "Numer. Solut. of Nonlinear Diff. Equ.", Wiley, 1966, 167-179. (Ch. V-VI).

A. Yu. LEPIN and A.D. MYSHKIS, Boundedness conditions for derivatives of bounded solutions of ordinary differential equations, Differential Équations 1 (1965) 989-991. (Ch. V).

J. LERAY and J. SCHAUDER, Topologie et équations fonctionnelles, Ann. Ecole Norm. Sup. (3) 51 (1934) 45-78.

D.C. LEWIS, On the role of first integrals in the perturbations of periodic solutions, Ann. Math. 63 (1956) 535-548.

J. LOCKER, An existence analysis for nonlinear equations in Hilbert space, Trans. Amer. Math. Soc. 128 (1967) 403-413.

J. LOCKER, An existence analysis for nonlinear boundary value problems, SIAM J. Appl. Math. 19 (1970) 199-207. (Ch. VI).

J. LOCKER, The method of least squares for boundary value problems, Trans. Amer. Math. Soc. 154 (1971) 57-68.

J. LOCKER, On constructing least squares solutions to two-point boundary value problems, Trans. Amer. Math. Soc. 203 (1975) 175-183.

A.M. LYAPUNOV, Sur les figures d'équilibre peu différentes des ellipsoïdes d'une masse liquide homogène dotée d'un mouvement de rotation, Zap. Akad. Nauk St. Petersbourg (1906) 1-225 ; (1908) 1-175 ; (1912) 1-228 ; (1914) 1-112.

R.J. MAGNUS, A generalization of multiplicity and the problem of bifurcation, Proc. London Math. Soc. (3) 32 ; (1976) 251-278. (Ch. X).

J. MAWHIN, Degré topologique et solutions périodiques des systèmes différentiels non linéaires, Bull. Soc. Roy. Sci. Liège 38 (1969) 308-398.

J. MAWHIN, Equations intégrales et solutions périodiques des systèmes différentiels non linéaires, Acad. Roy. Belgique Bull. Cl. Sci. (5) 55 (1969) 934-947.

J. MAWHIN, Equations fonctionnelles non linéaires et solutions périodiques, in "Equa-Diff 70", Marseille, 1970.

J. MAWHIN, Periodic solutions of nonlinear functional differential equations, J. Differential Equations 10 (1971) 240-261.

J. MAWHIN, "Equations non linéaires dans les espaces de Banach", Sémin. Math. Appl. Méc. Univ. Cath. Louvain, Rapport n° 39, Vander, Louvain, 1971.

J. MAWHIN, An extension of a theorem of A.C. Lazer on forced nonlinear oscillations, J. Math. Anal. Appl. 40 (1972) 20-29.

J. MAWHIN, Equivalence theorems for nonlinear operator equations and coincidence degree theory for some mappings in locally convex topological vector spaces, J. Differential Equations 12 (1972) 610-636.

J. MAWHIN, A further invariance property of coincidence degree in convex spaces, Ann. Soc. Sci. Bruxelles Ser. I 87 (1973) 51-57.

J. MAWHIN, "Degré de coïncidence et problèmes aux limites pour des équations différentielles ordinaires et fonctionnelles", Sémin. Math. Appl. Méc. Univ. Cath. Louvain, Rapport n° 64, Vander, Louvain, 1973.

J. MAWHIN, Problèmes aux limites du type de Neumann pour certaines équations différentielles ou aux dérivées partielles non linéaires, in "Equations différentielles et fonctionnelles non linéaires", Hermann, 1973, 124-134.

J. MAWHIN, The solvability of some operator equations with a quasibounded nonlinearity in normed spaces, J. Math. Anal. Appl. 45 (1974) 455-467.

J. MAWHIN, Periodic solutions of some vector retarded functional differential equations, J. Math. Anal. Appl. 45 (1974) 588-603.

J. MAWHIN, Boundary value problems for nonlinear second-order vector differential equations, J. Differential Equations 16 (1974) 257-269.

J. MAWHIN, L_2-estimates and periodic solutions of some nonlinear differential equations, Boll. Un. Mat. Ital. (4) 5 (1974) 343-354.

J. MAWHIN, "Nonlinear perturbations of Fredholm mappings in normed spaces and applications to differential equations", Univ. de Brasilia, Trabalho de Matematica n° 61, 1974.

J. MAWHIN, Nonlinear Functional Analysis and periodic solutions of ordinary differential equations, in "Diford 74", J.E. Purkyne Univ., Brno, 1974, 37-60.

J. MAWHIN, Periodic solutions of some perturbed differential equations, Boll. Un. Mat. Ital. (4) 11 (1975) 299-305.

J. MAWHIN, Recent results in coincidence theory for some mappings in normed spaces, Berichte Gesellschaft Math. Datenverarbeitung Bonn 103 (1975) 7-22.

J. MAWHIN, Recent results on periodic solutions of differential equations, in "International Conference on Differential Equations", Academic Press, New York, 1975, 537-556.

J. MAWHIN, Topology and nonlinear boundary value problem, in "Dynamical Systems : An International Symposium", vol. I, Academic Press, New York, 1976, 51-82.

J. MAWHIN, Nonlinear boundary value problems for ordinary differential equations : from Schauder theorem to stable homotopy, to appear in a volume dedicated to E. Rothe, Academic Press. (Ch. V, XII).

J. MAWHIN, Periodic solutions of nonlinear telegraph equations, Intern. Symp. Dynamical Systems, Gainesville, Florida, 1976, to appear. (Ch. VII).

J. MAWHIN, Stable homotopy and ordinary differential equations with nonlinear boundary conditions, to appear. (Ch. XII).

J. MAWHIN and C. MUNOZ, Application du degré topologique à l'estimation du nombre des solutions périodiques d'équations différentielles, Ann. Mat. Pura Appl. (4) 96 (1973) 1-19.

J. MAWHIN and K. SCHMITT, Rothe and Altman type coincidence theorems and applications to differential equations, to appear. (Ch. IV, VII).

V.B. MELAMED, On the computation of the index of the fixed point of a compact vector field (Russian), Dokl. Akad. Nauk SSSR 126 (1959) 591-594.

V.B. MELAMED, On the computation of the rotation of a compact vector field in the critical case (Russian) Sibirsk. Mat. Z. 2 (1961) 414-427.

Z. MIKOLAJSKA, Une remarque sur l'existence d'une solution périodique d'une équation différo-différentielle au deuxième membre croissant, Ann. Polon. Math. 18 (1966) 53-58. (Ch. IX).

M. MININNI, Coincidence degree and solvability of some nonlinear functional equations in normed spaces : a spectral approach, to appear. (Ch. VII).

C. MIRANDA, "Partial Differential Equations of Elliptic Type", Springer, Berlin, 1970.

R.D. MOYER, Second order differential equations of monotonic type, J. Differential Equations 2 (1966) 281-292. (Ch. V).

E. MUHAMADIEV, On the theory of periodic solutions of nonlinear systems of elliptic equations, Soviet Math. Dokl. 14 (1973) 226-229. (Ch. VIII).

K.K. MUKHERJEA, Cohomology theory for Banach manifolds, J. Math. Mech. 19 (1970) 731-744.

J.S. MULDOWNEY and D. WILLETT, An intermediate value property for operators with applications to integral and differential equations, Canadian J. Math. 26 (1974) 27-41. (Ch. V).

.S. MULDOWNEY and D. WILLETT, An elementary proof of the existence of solutions to second order nonlinear boundary value problems, SIAM J. Math. Anal. 5 (1974) 701-707. (Ch. V).

. MYJAK, Boundary value problems for nonlinear differential and difference equations of the second order, Zeszyty Nauk Univ. Jagiello. Prace Mat. Zeszyt 15 (1971) 113-123. (Ch. V, IX).

1. NAGUMO, Ueber die Differentialgleichung y" = f(x,y,y'), Proc. Phys.-Math. Soc. Japan (3) 19 (1937) 861-866.

1. NAGUMO, Ueber des Randwertproblem der nichtlinearen gewöhnlichen Differentialgleichungen zweiter Ordnung, Proc. Phys. Math. Soc. Japan 24 (1942) 845-851.

1. NAGUMO, Degree of mapping in convex linear topological spaces, Amer. J. Math. 73 (1951) 497-511.

. NECAS, On the range of nonlinear operators with linear asymptotes which are not invertible, Comment. Math. Univ. Carolinae 14 (1973) 63-72.

.. NIRENBERG, "Functional Analysis", New York University Lecture Notes, New York, 1960-61.

.. NIRENBERG, An application of generalized degree to a class of nonlinear problems, in "Troisième Colloque du C.B.R.M. d'analyse fonctionnelle", Vander, Louvain, 1971, 57-74.

.. NIRENBERG, Generalized degree and nonlinear problems, in "Contrib. Nonlinear Functional Analysis", Academic Press, 1971, 1-9. (Ch. XII).

.. NIRENBERG, "Topics in Nonlinear Functional Analysis", New York University Lecture Notes, 1973-1974.

.D. NUSSBAUM, The fixed point index for local condensing maps, An. Mat. Pura Appl. (4) 89 (1971) 217-258.

.D. NUSSBAUM, Degree theory for local condensing maps, J. Math. Anal. Appl. 37 (1972) 741-766.

. O'NEIL and J.W. THOMAS, On the equivalence of multiplicity and the generalized topological degree, Trans. Amer. Math. Soc. 167 (1972) 333-345.

.E. OSBORN and D. SATHER, Alternative problems for nonlinear equations, J. Differential Equations 17 (1975) 12-31.

.E. OSBORN and D. SATHER, Alternative problems and monotonicity, J. Differential Equations 18 (1975) 393-410. (Ch. II).

.V. PETRYSHYN, Fredholm alternative for nonlinear k-ball-contractive mappings with applications, J. Differential Equations 17 (1975) 82-95.

G. PRODI, Problemi di diramazione per equazioni funzionali, in "Atti Ottavo Congresso dell'Unione Mat. Ital.", Trieste, 1967, Bologna, 1968, 118-137. (Ch. X).

F. QUINN, Transversal approximation on Banach manifolds, Proc. Symp. Pure Math. 15 (1970) 213-227.

P.H. RABINOWITZ, Periodic solutions of nonlinear hyperbolic partial differential equations, Comm. Pure Appl. Math. 20 (1967) 145-206.

P.H. RABINOWITZ, Global theorems for nonlinear eigenvalue problems and applications, in "Contrib. to Nonlin. Functional Analysis", Academic Press, New York, 1971, 11-36. (Ch. X).

P.H. RABINOWITZ, Some global results for nonlinear eigenvalue problems, J. Functional Analysis 7 (1971) 487-513.

P.H. RABINOWITZ, A note on a nonlinear elliptic equation, Indiana Univ. Math. J. 22 (1972) 43-49.

P.H. RABINOWITZ, Some aspects of nonlinear eigenvalue problems, Rocky Mountain J. Math. 3 (1973) 161-202.

P.H. RABINOWITZ, "Théorie du degré topologique et applications à des problèmes aux limites non linéaires", Université Paris VI, Laboratoire Analyse Numérique, Paris, 1975.

P.H. RABINOWITZ, A survey of bifurcation theory, in "Dynamical Systems : An International Symposium", vol. I, Academic Press, New York, 1976, 83-96. (Ch. X).

R. REISSIG, "Periodic solutions - Forced oscillations", Study n° 6, Stability, Solid Mechanics Division, University of Waterloo.

R. REISSIG, On the existence of periodic solutions for a certain nonautonomous differential equation, Ann. Mat. Pura Appl. (4) 85 (1970) 235-240.

R. REISSIG, Note on a certain non-autonomous differential equation, Atti Accad. Naz. Lincei Rend. Cl. Sci. Fis. Mat. Natur. (8) 48 (1970) 484-486.

R. REISSIG, Periodic solutions of a nonlinear n-th order vector differential equation, Ann. Mat. Pura Appl. (4) 87 (1970) 111-124.

R. REISSIG, Über einen allgemeinen Typ erzwungener nichtlinearer Schwingungen zweiter Ordnung, Atti Accad. Naz. Lincei Rend. Cl. Sci. Fis. Mat. Natur. (8) 56 (1974) 297-302.

R. REISSIG, Extension of some results concerning the generalized Linéard equation, Ann. Mat. Pura Appl. (4) 104 (1975) 269-281.

R. REISSIG, Perturbation of a certain critical n-th order differential equation, Boll. Un. Mat. Ital. (4) 11 (1975), suppl-fasc. 3, 131-141.

R. REISSIG, Schwingungsätze für die verallgemeinerte Liénardsche Differentialgleichungen, Abh. Math. Sem. Univ. Hamburg 44 (1975) 45-51.

R. REISSIG, An extension of some oscillation theorems of Mawhin and Sedsiwy, Atti Accad. Naz. Lincei Rend. Cl. Sci. Fis. Mat. Natur., to appear.

E.H. ROTHE, On the Cesari index and the Browder-Petryshyn degree, in "Proceed. Symposium on Dynamical Systems", Gainesville, March 1976, to appear. (Ch. VI and XI).

N. ROUCHE and J. MAWHIN, "Equations différentielles ordinaires", 2 volumes, Masson, Paris, 1973.

B.N. SADOVSKII, Application of topological methods in the theory of periodic solutions of nonlinear differential-operator equations of neutral type, Soviet Math. Dokl. 12 (1971) 1543-1547.

B.N. SADOVSKII, Limit-compact and condensing operators, Russian Math. Surveys 27 (1972) 85-156.

D.A. SANCHEZ, An iteration scheme for boundary value/alternative problems, MRC Technical Summary Report n° 1412, January 1974. (Ch. II).

D.A. SANCHEZ, An iteration scheme for Hilbert space boundary value problems, Bol. Un. Mat. Ital. (4) 11 (1975) 1-9.

D. SATHER, Branching of solutions of nonlinear equations, Rocky Mountain J. Math. 3 (1973) 203-250. (Ch. II).

D.H. SATTINGER, Stability of bifurcating solutions by Leray-Schauder degree, Arch. Rat. Mech. Anal. 43 (1971) 154-166. (Ch. X).

D.H. SATTINGER, "Topics in Stability and Bifurcation Theory", Springer, Lecture Notes in Mathematics n° 309, Berlin, 1973. (Ch. X).

M. SCHECHTER, A nonlinear elliptic boundary value problem, Ann. Scuola Norm. Sup. Pisa (3) 27 (1973) 707-716.

E. SCHMIDT, Zur Theorie des linearen und nichtlinearen Integralgleichungen. 3 Teil. Uber die Auflösung der nichtlinearen Integralgleichungen und die Verzeigung ihrer Lösungen, Math. Ann. 65 (1908) 370-399.

K. SCHMITT, Periodic solutions of nonlinear second order differential equations, Math. Z. 98 (1967) 200-207.

K. SCHMITT, Periodic solutions of systems of second order differential equations, J. Differential Equations 11 (1972) 180-192.

K. SCHMITT, Intermediate value theorems for periodic functional differential equations, in "Equations différentielles et fonctionnelles non linéaires", Hermann, Paris, 1973, 65-78.

K. SCHMITT, Randwertaufgaben für gewöhnliche Differentialgleichungen, in "Proc. Steiermark. Math. Symposium", Graz, 1973, VII, 1-55.

K. SCHMITT, Applications of variational equations to ordinary and partial differential equations - Multiple solutions of boundary value problems, J. Differential Equations 17 (1975) 154-186. (Ch. V).

K. SCHMITT and R. THOMPSON, Boundary value problems for infinite systems of second-order differential equations, J. Differential Equations 18 (1975) 277-295. (Ch. V).

K.R. SCHNEIDER, Uber eine Integralgleichungsmethode in der Theorie autonomer Schwingungen und ihre Anwendung auf Bifurkationsprobleme, Math. Nachr. 69 (1975) 191-204. (Ch. X).

K.R. SCHNEIDER, Investigations of integral equations in the theory of nonlinear oscillations, Abh. Akad. Wiss. Deutsche Demokr. Rep., to appear. (Ch. II, V, IX).

K.R. SCHNEIDER and B. WEGNER, Asymptotic behavior of the period of a closed trajectory contracting to a singular point, to appear. (Ch. X).

K.W. SCHRADER, Solutions of second order ordinary differential equations, J. Differential Equations 4 (1968) 510-518. (Ch. V).

J. SCHRÖDER, Upper and lower bounds for the solutions of two-point boundary value problems, Inst. Fluid Dynam. Appl. Math., Univ. Maryland, Techn. Note BN-797, July 1974. (Ch. V).

J. SCHRÖDER, Upper and lower bounds for solutions of generalized two-point boundary value problems, Numer. Math. 23 (1975) 433-457. (Ch. V).

J.T. SCHWARTZ, "Nonlinear Functional Analysis", Gordon and Breach, New York, 1969.

S. SEDSIWY, On periodic solutions of a certain third-order non-linear differential equation, Ann. Polon. Math. 17 (1965) 147-154.

S. SEDSIWY, Asymptotic properties of solutions of nonlinear differential equations, Zeszyty Nauk Univ. Jagiell. N° 131 (1966) 69-80.

S. SEDSIWY, Asymptotic properties of solutions of a certain n-th order vector differential equation, Atti Accad. Naz. Lincei Rend. Cl. Sci. Fis. Mat. Natur. (8) 47 (1969) 472-475.

S. SEDSIWY, Periodic solutions of a system of nonlinear differential equations, Proc. Amer. Math. Soc. 48 (1975) 328-336.

T. SHIMIZU, Analytic operations and analytic operational equations, Math. Japon. 1 (1948) 36-40.

S. SMALE, An infinite dimensional version of Sard's theorem, Amer. J. Math. 87 (1965) 861-866.

.. SMITALOVA, Periodic solutions of a second order nonlinear differential equation, Acta Fac. Rer. Nat. Univ. Comenian, Math. 27 (1972) 19-25. (Ch. IX).

. SOVA, Abstract semilinear equations with small nonlinearities, Comment. Math. Univ. Carolinae 12 (1971) 785-805.

. STAKGOLD, Branching of solutions of nonlinear equations, SIAM Review 13 (1971) 289-392. (Ch. X).

. STAMPACCHIA, Problemi al contorno ellittici, con dati discontinui, dotati di soluzioni hölderiane, Ann. Mat. Pura Appl. (4) 51 (1960) 1-37.

. STOKES, Invariant subspaces, symmetry and alternative problems, Bull. Inst. Math. Acad. Sinica 3 (1975) 7-14. (Ch. II).

. STRASBERG, "La recherche de solutions périodiques d'équations différentielles non linéaires", Ph.D. Thesis, Univ. Libre de Bruxelles, 1975.

.V. STRYGIN, A theorem concerning the existence of periodic solutions of systems of differential equations with delayed arguments, Math. Notes 8 (1970) 600-602.

.A. STUART, Some bifurcation theory for k-set contractions, Proc. London Math. Soc. (3) 27 (1973) 531-550.

.A. STUART and J.F. TOLAND, A global result applicable to nonlinear Steklov problems, J. Differential Equations 15 (1974) 247-268.

. STÜBEN, Einschliessunaussagen bei Differentialgleichungen mit periodische Randbedingungen, Math. Inst. Univ. Köln, Report 75-6, April 1975. (Ch. V).

.S. SVARC, The homotopic topology of Banach space, Soviet Math. Dokl. 5 (1964) 57-59.

. SWEET, An alternative method based on approximate solutions, Math. Systems Theory 4 (1970) 306-315.

. SWEET, Projection operator for the alternative problem, Univ. of Maryland, Technical Report TR 72-8, 1972.

. SWEET, Characterizing finite projections for functional equations, to appear. (Ch. VI).

.W. THOMAS, The multiplicity of an operator is a special case of the topological degree for k-set contractions, Duke Math. J. 40 (1973) 233-240.

.W. THOMAS, On a lower bound to the number of solutions to a nonlinear operator equation, Scripta Math. to appear.

.J. TROMBA, Stable homotopy groups of spheres and non-linear P.D.E., to appear.

.M. VAINBERG and P.G. AIZENGENDLER, The theory and methods of investigation of branch points of solutions, in "Progress in Math." 2, Plenum Press, New York, 1968, 1-72.

M.M. VAINBERG and V.A. TRENOGIN, The method of Lyapunov and Schmidt in the theory of nonlinear differential equations and their further development, Russian Math. Surveys 17 (1962) 1-60.

M.M. VAINBERG and V.A. TRENOGIN, "Theory of Branching of Solutions of Non-linear Equations", Noordhoff, Leyden, 1974. (Ch. II).

G.M. VAINIKKO and B.N. SADOVSKII, On the rotation of condensing vector fields (Russián), Probl. Matem. Analiza Slozhn. Sist. n° 2, Voronezh (1968) 84-88.

A. VANDERBAUWHEDE, Alternative problems and symmetry, to appear. (Ch. II).

A. VANDERBAUWHEDE, Alternative problems and invariant subspaces, to appear, (Ch. II).

L.M. VASERMAN and A.Ya. LEPIN, Generalized Bernstein conditions for nonlinear systems of ordinary differential equations, Differential Equations 5 (1969) 816-821. (Ch. V).

G. VILLARI, Soluzioni periodiche di una classe di equazioni differenziali del terz'ordine quasi lineari, Ann. Mat. Pura Appl. (4) 73 (1966) 103-110.

N. VORNICESCU, A supra existentei unei solutii periodice pentru ecuatia oscillatiilat neliniare fortate, St. Cerc. Mat. 22 (1970) 683-689.

S.A. WILLIAMS, A connection between the Cesari and Leray-Schauder methods, Michigan Math. J. 15 (1968) 441-448.

S.A. WILLIAMS, A sharp sufficient condition for solution of a nonlinear elliptic boundary value problem, J. Differential Equations 8 (1970) 580-586.

P.P. ZABREIKO and S.O. STRYGINA, Periodic solutions of evolutionary equations, Math. Notes 9 (1971) 378-384.

E. ZEIDLER, Existenz, Eindeutigkeit, Eigenschaften und Anwendungen des Abbildungsgrades in R^n, in "Theory of Nonlinear Operators. Proceedings of a Summer School", Akademie-Verlag, Berlin, 1974, 259-311. (Ch. III, IV).

INDEX

Vol. 399: Functional Analysis and its Applications. Proceedings 1973. Edited by H. G. Garnir, K. R. Unni and J. H. Williamson. , 584 pages. 1974.

ol. 400: A Crash Course on Kleinian Groups. Proceedings 1974. dited by L. Bers and I. Kra. VII, 130 pages. 1974.

ol. 401: M. F. Atiyah, Elliptic Operators and Compact Groups. 93 pages. 1974.

ol. 402: M. Waldschmidt, Nombres Transcendants. VIII, 277 ges. 1974.

l. 403: Combinatorial Mathematics. Proceedings 1972. Edited D. A. Holton. VIII, 148 pages. 1974.

l. 404: Théorie du Potentiel et Analyse Harmonique. Edité par Faraut. V, 245 pages. 1974.

. 405: K. J. Devlin and H. Johnsbråten, The Souslin Problem. , 132 pages. 1974.

. 406: Graphs and Combinatorics. Proceedings 1973. Edited R. A. Bari and F. Harary. VIII, 355 pages. 1974.

. 407: P. Berthelot, Cohomologie Cristalline des Schémas de racteristique p > o. II, 604 pages. 1974.

408: J. Wermer, Potential Theory. VIII, 146 pages. 1974.

409: Fonctions de Plusieurs Variables Complexes, Séminaire nçois Norguet 1970–1973. XIII, 612 pages. 1974.

410: Séminaire Pierre Lelong (Analyse) Année 1972–1973. 81 pages. 1974.

411: Hypergraph Seminar. Ohio State University, 1972. ed by C. Berge and D. Ray-Chaudhuri. IX, 287 pages. 1974.

412: Classification of Algebraic Varieties and Compact nplex Manifolds. Proceedings 1974. Edited by H. Popp. V, pages. 1974.

413: M. Bruneau, Variation Totale d'une Fonction. XIV, 332 es. 1974.

414: T. Kambayashi, M. Miyanishi and M. Takeuchi, Unint Algebraic Groups. VI, 165 pages. 1974.

415: Ordinary and Partial Differential Equations. Proceedings . XVII, 447 pages. 1974.

416: M. E. Taylor, Pseudo Differential Operators. IV, 155 s. 1974.

417: H. H. Keller, Differential Calculus in Locally Convex es. XVI, 131 pages. 1974.

418: Localization in Group Theory and Homotopy Theory Related Topics. Battelle Seattle 1974 Seminar. Edited by P. J. n. VI, 172 pages 1974.

419: Topics in Analysis. Proceedings 1970. Edited by O. E. , I. S. Louhivaara, and R. H. Nevanlinna. XIII, 392 pages. 1974.

20: Category Seminar. Proceedings 1972/73. Edited by G. M. VI, 375 pages. 1974.

421: V. Poénaru, Groupes Discrets. VI, 216 pages. 1974.

22: J.-M. Lemaire, Algèbres Connexes et Homologie des ces de Lacets. XIV, 133 pages. 1974.

23: S. S. Abhyankar and A. M. Sathaye, Geometric Theory gebraic Space Curves. XIV, 302 pages. 1974.

424: L. Weiss and J. Wolfowitz, Maximum Probability ators and Related Topics. V, 106 pages. 1974.

25: P. R. Chernoff and J. E. Marsden, Properties of Infinite nsional Hamiltonian Systems. IV, 160 pages. 1974.

26: M. L. Silverstein, Symmetric Markov Processes. X, 287 . 1974.

27: H. Omori, Infinite Dimensional Lie Transformation s. XII, 149 pages. 1974.

28: Algebraic and Geometrical Methods in Topology, Prongs 1973. Edited by L. F. McAuley. XI, 280 pages. 1974.

Vol. 429: L. Cohn, Analytic Theory of the Harish-Chandra C-Function. III, 154 pages. 1974.

Vol. 430: Constructive and Computational Methods for Differential and Integral Equations. Proceedings 1974. Edited by D. L. Colton and R. P. Gilbert. VII, 476 pages. 1974.

Vol. 431: Séminaire Bourbaki – vol. 1973/74. Exposés 436–452. IV, 347 pages. 1975.

Vol. 432: R. P. Pflug, Holomorphiegebiete, pseudokonvexe Gebiete und das Levi-Problem. VI, 210 Seiten. 1975.

Vol. 433: W. G. Faris, Self-Adjoint Operators. VII, 115 pages. 1975.

Vol. 434: P. Brenner, V. Thomée, and L. B. Wahlbin, Besov Spaces and Applications to Difference Methods for Initial Value Problems. II, 154 pages. 1975.

Vol. 435: C. F. Dunkl and D. E. Ramirez, Representations of Commutative Semitopological Semigroups. VI, 181 pages. 1975.

Vol. 436: L. Auslander and R. Tolimieri, Abelian Harmonic Analysis, Theta Functions and Function Algebras on a Nilmanifold. V, 99 pages. 1975.

Vol. 437: D. W. Masser, Elliptic Functions and Transcendence. XIV, 143 pages. 1975.

Vol. 438: Geometric Topology. Proceedings 1974. Edited by L. C. Glaser and T. B. Rushing. X, 459 pages. 1975.

Vol. 439: K. Ueno, Classification Theory of Algebraic Varieties and Compact Complex Spaces. XIX, 278 pages. 1975

Vol. 440: R. K. Getoor, Markov Processes: Ray Processes and Right Processes. V, 118 pages. 1975.

Vol. 441: N. Jacobson, PI-Algebras. An Introduction. V, 115 pages. 1975.

Vol. 442: C. H. Wilcox, Scattering Theory for the d'Alembert Equation in Exterior Domains. III, 184 pages. 1975.

Vol. 443: M. Lazard, Commutative Formal Groups. II, 236 pages. 1975.

Vol. 444: F. van Oystaeyen, Prime Spectra in Non-Commutative Algebra. V, 128 pages. 1975.

Vol. 445: Model Theory and Topoi. Edited by F. W. Lawvere, C. Maurer, and G. C. Wraith. III, 354 pages. 1975.

Vol. 446: Partial Differential Equations and Related Topics. Proceedings 1974. Edited by J. A. Goldstein. IV, 389 pages. 1975.

Vol. 447: S. Toledo, Tableau Systems for First Order Number Theory and Certain Higher Order Theories. III, 339 pages. 1975.

Vol. 448: Spectral Theory and Differential Equations. Proceedings 1974. Edited by W. N. Everitt. XII, 321 pages. 1975.

Vol. 449: Hyperfunctions and Theoretical Physics. Proceedings 1973. Edited by F. Pham. IV, 218 pages. 1975.

Vol. 450: Algebra and Logic. Proceedings 1974. Edited by J. N. Crossley. VIII, 307 pages. 1975.

Vol. 451: Probabilistic Methods in Differential Equations. Proceedings 1974. Edited by M. A. Pinsky. VII, 162 pages. 1975.

Vol. 452: Combinatorial Mathematics III. Proceedings 1974. Edited by Anne Penfold Street and W. D. Wallis. IX, 233 pages. 1975.

Vol. 453: Logic Colloquium. Symposium on Logic Held at Boston, 1972–73. Edited by R. Parikh. IV, 251 pages. 1975.

Vol. 454: J. Hirschfeld and W. H. Wheeler, Forcing, Arithmetic, Division Rings. VII, 266 pages. 1975.

Vol. 455: H. Kraft, Kommutative algebraische Gruppen und Ringe. III, 163 Seiten. 1975.

Vol. 456: R. M. Fossum, P. A. Griffith, and I. Reiten, Trivial Extensions of Abelian Categories. Homological Algebra of Trivial Extensions of Abelian Categories with Applications to Ring Theory. XI, 122 pages. 1975.

Vol. 457: Fractional Calculus and Its Applications. Proceedings 1974. Edited by B. Ross. VI, 381 pages. 1975.

Vol. 458: P. Walters, Ergodic Theory – Introductory Lectures. VI, 198 pages. 1975.

Vol. 459: Fourier Integral Operators and Partial Differential Equations. Proceedings 1974. Edited by J. Chazarain. VI, 372 pages. 1975.

Vol. 460: O. Loos, Jordan Pairs. XVI, 218 pages. 1975.

Vol. 461: Computational Mechanics. Proceedings 1974. Edited by J. T. Oden. VII, 328 pages. 1975.

Vol. 462: P. Gérardin, Construction de Séries Discrètes p-adiques. »Sur les séries discrètes non ramifiées des groupes réductifs déployés p-adiques«. III, 180 pages. 1975.

Vol. 463: H.-H. Kuo, Gaussian Measures in Banach Spaces. VI, 224 pages. 1975.

Vol. 464: C. Rockland, Hypoellipticity and Eigenvalue Asymptotics. III, 171 pages. 1975.

Vol. 465: Séminaire de Probabilités IX. Proceedings 1973/74. Edité par P. A. Meyer. IV, 589 pages. 1975.

Vol. 466: Non-Commutative Harmonic Analysis. Proceedings 1974. Edited by J. Carmona, J. Dixmier and M. Vergne. VI, 231 pages. 1975.

Vol. 467: M. R. Essén, The Cos $\pi\lambda$ Theorem. With a paper by Christer Borell. VII, 112 pages. 1975.

Vol. 468: Dynamical Systems – Warwick 1974. Proceedings 1973/74. Edited by A. Manning. X, 405 pages. 1975.

Vol. 469: E. Binz, Continuous Convergence on C(X). IX, 140 pages. 1975.

Vol. 470: R. Bowen, Equilibrium States and the Ergodic Theory of Anosov Diffeomorphisms. III, 108 pages. 1975.

Vol. 471: R. S. Hamilton, Harmonic Maps of Manifolds with Boundary. III, 168 pages. 1975.

Vol. 472: Probability-Winter School. Proceedings 1975. Edited by Z. Ciesielski, K. Urbanik, and W. A. Woyczyński. VI, 283 pages. 1975.

Vol. 473: D. Burghelea, R. Lashof, and M. Rothenberg, Groups of Automorphisms of Manifolds. (with an appendix by E. Pedersen) VII, 156 pages. 1975.

Vol. 474: Séminaire Pierre Lelong (Analyse) Année 1973/74. Edité par P. Lelong. VI, 182 pages. 1975.

Vol. 475: Répartition Modulo 1. Actes du Colloque de Marseille-Luminy, 4 au 7 Juin 1974. Edité par G. Rauzy. V, 258 pages. 1975. 1975.

Vol. 476: Modular Functions of One Variable IV. Proceedings 1972. Edited by B. J. Birch and W. Kuyk. V, 151 pages. 1975.

Vol. 477: Optimization and Optimal Control. Proceedings 1974. Edited by R. Bulirsch, W. Oettli, and J. Stoer. VII, 294 pages. 1975.

Vol. 478: G. Schober, Univalent Functions – Selected Topics. V, 200 pages. 1975.

Vol. 479: S. D. Fisher and J. W. Jerome, Minimum Norm Extremals in Function Spaces. With Applications to Classical and Modern Analysis. VIII, 209 pages. 1975.

Vol. 480: X. M. Fernique, J. P. Conze et J. Gani, Ecole d'Eté de Probabilités de Saint-Flour IV-1974. Edité par P.-L. Hennequin. XI, 293 pages. 1975.

Vol. 481: M. de Guzmán, Differentiation of Integrals in R^n. XII, 226 pages. 1975.

Vol. 482: Fonctions de Plusieurs Variables Complexes II. Séminaire François Norguet 1974–1975. IX, 367 pages. 1975.

Vol. 483: R. D. M. Accola, Riemann Surfaces, Theta Functions, and Abelian Automorphisms Groups. III, 105 pages. 1975.

Vol. 484: Differential Topology and Geometry. Proceedings 1974. Edited by G. P. Joubert, R. P. Moussu, and R. H. Roussarie. IX, 287 pages. 1975.

Vol. 485: J. Diestel, Geometry of Banach Spaces – Selected Topics. XI, 282 pages. 1975.

Vol. 486: S. Stratila and D. Voiculescu, Representations of AF-Algebras and of the Group U (∞). IX, 169 pages. 1975.

Vol. 487: H. M. Reimann und T. Rychener, Funktionen beschränkter mittlerer Oszillation. VI, 141 Seiten. 1975.

Vol. 488: Representations of Algebras, Ottawa 1974. Proceedings 1974. Edited by V. Dlab and P. Gabriel. XII, 378 pages. 1975.

Vol. 489: J. Bair and R. Fourneau, Etude Géométrique des Espaces Vectoriels. Une Introduction. VII, 185 pages. 1975.

Vol. 490: The Geometry of Metric and Linear Spaces. Proceeding 1974. Edited by L. M. Kelly. X, 244 pages. 1975.

Vol. 491: K. A. Broughan, Invariants for Real-Generated Unifor Topological and Algebraic Categories. X, 197 pages. 1975.

Vol. 492: Infinitary Logic: In Memoriam Carol Karp. Edited by D. Kueker. VI, 206 pages. 1975.

Vol. 493: F. W. Kamber and P. Tondeur, Foliated Bundles a Characteristic Classes. XIII, 208 pages. 1975.

Vol. 494: A Cornea and G. Licea. Order and Potential Resolv Families of Kernels. IV, 154 pages. 1975.

Vol. 495: A. Kerber, Representations of Permutation Groups II. 175 pages. 1975.

Vol. 496: L. H. Hodgkin and V. P. Snaith, Topics in K-Theory. T Independent Contributions. III, 294 pages. 1975.

Vol. 497: Analyse Harmonique sur les Groupes de Lie. Proceedi 1973–75. Edité par P. Eymard et al. VI, 710 pages. 1975.

Vol. 498: Model Theory and Algebra. A Memorial Tribute Abraham Robinson. Edited by D. H. Saracino and V. B. Weispfenn X, 463 pages. 1975.

Vol. 499: Logic Conference, Kiel 1974. Proceedings. Edited G. H. Müller, A. Oberschelp, and K. Potthoff. V, 651 pages 19

Vol. 500: Proof Theory Symposion, Kiel 1974. Proceedings. Edite J. Diller and G. H. Müller. VIII, 383 pages. 1975.

Vol. 501: Spline Functions, Karlsruhe 1975. Proceedings. Edite K. Böhmer, G. Meinardus, and W. Schempp. VI, 421 pages. 1976.

Vol. 502: János Galambos, Representations of Real Number Infinite Series. VI, 146 pages. 1976.

Vol. 503: Applications of Methods of Functional Analysis to Probl in Mechanics. Proceedings 1975. Edited by P. Germain an Nayroles. XIX, 531 pages. 1976.

Vol. 504: S. Lang and H. F. Trotter, Frobenius Distribution GL_2-Extensions. III, 274 pages. 1976.

Vol. 505: Advances in Complex Function Theory. Proceed 1973/74. Edited by W. E. Kirwan and L. Zalcman. VIII, 203 pa 1976.

Vol. 506: Numerical Analysis, Dundee 1975. Proceedings. Ed by G. A. Watson. X, 201 pages. 1976.

Vol. 507: M. C. Reed, Abstract Non-Linear Wave Equations 128 pages. 1976.

Vol. 508: E. Seneta, Regularly Varying Functions. V, 112 pages. 19

Vol. 509: D. E. Blair, Contact Manifolds in Riemannian Geon VI, 146 pages. 1976.

Vol. 510: V. Poènaru, Singularités C^∞ en Présence de Sym V, 174 pages. 1976.

Vol. 511: Séminaire de Probabilités X. Proceedings 1974/75. par P. A. Meyer. VI, 593 pages. 1976.

Vol. 512: Spaces of Analytic Functions, Kristiansand, Norway Proceedings. Edited by O. B. Bekken, B. K. Øksendal, and A. VIII, 204 pages. 1976.

Vol. 513: R. B. Warfield, Jr. Nilpotent Groups. VIII, 115 pages.

Vol. 514: Séminaire Bourbaki vol. 1974/75. Exposés 453 – 4 276 pages. 1976.

Vol. 515: Bäcklund Transformations. Nashville, Tennessee Proceedings. Edited by R. M. Miura. VIII, 295 pages. 1976.